English for Chemistry
化学专业英语

主　编　谭明雄　黄国保　张　培
副主编　杨　燕　曾玉凤　李海斌
　　　　覃其品　钟玉君

西南交通大学出版社
·成　都·

图书在版编目（CIP）数据

化学专业英语 = English for Chemistry / 谭明雄，黄国保，张培主编. —成都：西南交通大学出版社，2023.8
　ISBN 978-7-5643-9388-5

Ⅰ.①化… Ⅱ.①谭…②黄…③张… Ⅲ.①化学–英语–高等学校–教材 Ⅳ.①O6

中国国家版本馆 CIP 数据核字（2023）第 126597 号

English for Chemistry
化学专业英语

主　编 / 谭明雄　黄国保　张　培　　责任编辑 / 牛　君
　　　　　　　　　　　　　　　　　　封面设计 / 何东琳设计工作室

西南交通大学出版社出版发行
（四川省成都市金牛区二环路北一段 111 号西南交通大学创新大厦 21 楼　610031）
发行部电话：028-87600564　　028-87600533
网址：http://www.xnjdcbs.com
印刷：四川煤田地质制图印务有限责任公司

成品尺寸　185 mm × 260 mm
印张　6.75　　字数　174 千
版次　2023 年 8 月第 1 版　　印次　2023 年 8 月第 1 次

书号　ISBN 978-7-5643-9388-5
定价　22.00 元

课件咨询电话：028-81435775
图书如有印装质量问题　本社负责退换
版权所有　盗版必究　举报电话：028-87600562

前言 Preface

随着化学、化工、材料等领域的快速发展,新技术、新工艺、新材料不断出现。对于相关专业的学生来说,不管在学习还是以后的科研工作中,化学专业英语的学习对精准理解专业前沿知识、掌握专业技能、拓宽专业视野都具有重要作用。

本书由玉林师范学院化学、材料、化工等专业的教授、博士编写。编写老师积累了多年化学、化工、材料专业英语的授课经验,根据化学、化工、材料等专业领域从事科学研究、科技服务等对专业英语的要求,结合学科发展需要,内容力求实用、新颖,体现科技前沿和实际应用,体现专业词汇构词法及专业文献语法特点,能把英语的基本知识和化学专业知识紧密地联结起来,强化专业知识和基础英语的实践应用。

本书内容分为四部分,涉及无机化学、有机化学、材料化学、化工过程等相关基础知识,素材来源于化学、化工、材料类英文网站、专著、杂志以及相关专业教学用书,选材面广,专业性强,难易适中,循序渐进,突出专业英语语法特点、翻译技巧、专业词汇构词法,如常用化学化工词头、词尾、缩写词和各种有机化合物词汇的构成和书写,化学方程式的英文描述方式与技巧等,内容精简、通俗易懂,有效地避免了冗长乏味的感觉。帮助学生了解专业词汇和专业语法特点,掌握专业英语知识的使用和实践。

本书内容丰富,取材有代表性和前沿性,适合作为高等院校化学、化工、材料等专业本科学生专业英语的学习用书,也可作为从事化学、化工、材料领域的教学、科研和工程技术人员的参考用书。

本书由谭明雄、黄国保、张培主编,多位教师参与编写。其中第一部分元素和无机化合物由谭明雄和黄国保老师编写,第二部分各类有机化合物由谭明雄和杨燕老师编写,第三部分材料化学基本知识由张培老师编写,第四部分典型化工过程由曾玉凤和李海斌老师编写,覃其品和钟玉君老师负责部分校对工作。

由于编者水平有限,书中疏漏与不妥之处在所难免,恳请读者和同仁提出宝贵意见并批评指正。

编 者

2022 年 11 月

Contents

Part 1 Elements and Inorganic Compounds ··············· 1
- Lesson 1 Elements and Symbols ··············· 3
- Lesson 2 The Periodic Table ··············· 6
- Lesson 3 The Nonmetal Elements ··············· 10
- Lesson 4 The Metallic Elements ··············· 14
- Lesson 5 Ionic and Molecular Compounds ··············· 21
- Lesson 6 Acids, Bases and Equilibrium ··············· 28
- Lesson 7 Reactions of Acids and Bases ··············· 35

Part 2 Various Kinds of Organic Compounds ··············· 39
- Lesson 8 Alkanes and Cycloalkanes ··············· 41
- Lesson 9 Alkenes and Alkynes ··············· 47
- Lesson 10 Alcohols, Aldehydes and Ketones ··············· 51
- Lesson 11 Carboxylic Acids, Esters, Amines and Amides ··············· 57

Part 3 Basic Knowledge of Materials Chemistry ··············· 63
- Lesson 12 Characteristics and Applications of Metals ··············· 65
- Lesson 13 Characteristics and Applications of Ceramics ··············· 72
- Lesson 14 Characteristics and Applications of Polymers ··············· 75
- Lesson 15 Characteristics and Applications of Composites ··············· 79

Part 4 Typical Chemical Process ··············· 83
- Lesson 16 Operations in Chemical Engineering ··············· 85
- Lesson 17 Distillation ··············· 90
- Lesson 18 Wet-Chemical Synthesis of Inorganic Nanocrystals ··············· 96

References ··············· 101

Part 1

Elements and Inorganic Compounds

Part 1　Elements and Inorganic Compounds

Lesson 1　Elements and Symbols

All matter is composed of elements, of which there are 118 different kinds. Of these, 88 elements occur naturally and make up all the substances in our world. Many elements are already familiar to you. Perhaps you use aluminum in the form of foil or drink soft drinks from aluminum cans. You may have a ring or necklace made of gold, silver, or perhaps platinum.

If you play tennis or golf, then you may have noticed that your racket or clubs may be made from the elements titanium or carbon. In our bodies, calcium and phosphorus form the structure of bones and teeth, iron and copper are needed in the formation of red blood cells, and iodine is required for the proper functioning of the thyroid.

The correct amounts of certain elements are crucial to the proper growth and function of the body. Low levels of iron can lead to anemia, while lack of iodine can cause hypothyroidism and goiter. Some elements known as microminerals, such as chromium, cobalt, and selenium, are needed in our bodies in very small amounts. Laboratory tests are used to confirm that these elements are within normal ranges in our bodies.

Elements are pure substances from which all other things are built. Elements cannot be broken down into simpler substances. Chemical symbols are one- or two-letter abbreviations for the names of the elements. Only the first letter of an element's symbol is capitalized. If the symbol has a second letter, it is lowercase so that we know when a different element is indicated. If two letters are capitalized, they represent the symbols of two different elements. For example, the element cobalt has the symbol Co. However, the two capital letters CO specify two elements, carbon (C) and oxygen (O).

Although most of the symbols use letters from the current names, some are derived from their ancient names. For example, Na, the symbol for sodium, comes from the Latin word *natrium*. The symbol for iron, Fe, is derived from the Latin name *ferrum*. Table 1.1 lists the names and symbols of some common elements.

Table 1.1 Names and Symbols of Some Common Elements

Name*	Symbol	Name*	Symbol	Name*	Symbol
Aluminum	Al	Gold(*aurum*)	Au	Oxygen	O
Argon	Ar	Helium	He	Phosphorus	P
Arsenic	As	Hydrogen	H	Platinum	Pt
Barium	Ba	Iodine	I	Potassium(*kalium*)	K
Boron	B	Iron(*ferrum*)	Fe	Radium	Ra
Bromine	Br	Lead(*plumbum*)	Pb	Silicon	Si
Cadmium	Cd	Lithium	Li	Silver(*argentums*)	Ag
Calcium	Ca	Magnesium	Mg	Sodium(*natrium*)	Na
Carbon	C	Manganese	Mn	Strontium	Sr
Chlorine	Cl	Mercury(*hydrargyrum*)	Hg	Sulfur	S
Chromium	Cr	Neon	Ne	Tin(*stannum*)	Sn
Cobalt	Co	Nickel	Ni	Titanium	Ti
Copper(*cuprum*)	Cu	Nitrogen	N	Uranium	U
Fluorine	F			Zinc	Zn

* Names given in parentheses are ancient Latin or Greek words from which the symbols are derived.

Vocabulary

matter ['mætə(r)]　*n.* 物质，问题
compose [kəm'pəuz]　*v.* 组成
titanium [ti'teɪniəm]　*n.* 钛
element ['elimənt]　*n.* 元素
substance ['sʌbstəns]　*n.* 物质
aluminum [ə'lju:minəm]　*n.* 铝
foil [fɔil]　*n.* 箔
silver ['silvə]　*n.* 银
platinum ['plætinəm]　*n.* 白金
Racket ['rækit]　*n.* 球拍
club [klʌb]　*n.* 俱乐部，球杆
carbon ['kɑ:bən]　*n.* 碳
calcium ['kælsiəm]　*n.* 钙
phosphorus ['fɔsfərəs]　*n.* 磷；磷光体
iron ['aiən]　*n.* 铁器，铁制品
copper ['kɒpə]　*n.* 铜

iodine ['aɪədiːn]   n. 碘
thyroid ['θaɪrɔid]   n. 甲状腺；甲状软骨
hypothyroidism [ˌhaipəu'θairɔidizəm]   n. 甲状腺机能减退
goiter ['gɔitə]   n. 甲状腺肿
chromium ['krəumiəm]   n. 铬
cobalt ['kəubɔːlt]   n. 钴
selenium [sə'liːniəm]   n. 硒
laboratory [lə'bɔrətri]   n. 实验室；实验课
element ['elimənt]   n. 元素
substances ['sʌbstənsɪz]   n. 物质；化学物质
abbreviation [əˌbriːvi'eiʃn]   n. 缩写；省略
symbol ['simb(ə)l]   n. 符号；象征
capitalize ['kæpitəlaiz]   v. 把……大写
lowercase ['ləuəˌkeis]   v. 英文小写；全部小写
indicate ['indikeit]   v. 指示；指出；表明
represent [ˌrepri'zent]   v. 代表；表现；体现
chemical ['kemikl]   n. 化学品；化学制品
            adj. 与化学有关的；化学的
simpler ['simplə]   adj. 更简单的；更轻便的；更单纯的
ferrum ['ferəm]   n. 铁
sodium ['səudiəm]   n. 钠
oxygen ['ɔksidʒən]   n. 氧；氧气
specify ['spesifai]   v. 具体说明；明确规定；详述；详列

Lesson 2 The Periodic Table

As more elements were discovered, it became necessary to organize them into some type of classification system. By the late 1800s, scientists recognized that certain elements looked alike and behaved much the same way. In 1872, a Russian chemist, Dmitri Mendeleev, arranged the 60 elements known at that time into groups with similar properties and placed them in order of increasing atomic masses. Today, this arrangement of 118 elements is known as the periodic table (see Figure 2.1).

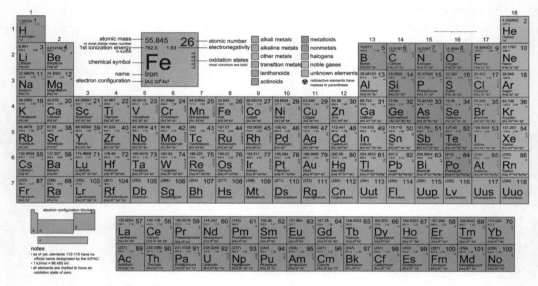

Figure 2.1 Periodic Table of the Elements

Names of Groups

Several groups in the periodic table have special names. Group ⅠA—lithium (Li), sodium (Na), potassium (K), rubidium (Rb), cesium (Cs), and francium (Fr)—are a family of elements known as the alkali metals. The elements within this group are soft, shiny metals that are good conductors of heat and electricity and have relatively low melting points. Alkali metals react vigorously with water and form white products when they combine with oxygen.

Although hydrogen (H) is at the top of Group ⅠA, it is not an alkali metal and has very different properties than the rest of the elements in this group. Thus hydrogen is not included in the alkali metals.

The alkaline earth metals are found in Group ⅡA. They include the elements beryllium (Be), magnesium (Mg), calcium (Ca), strontium (Sr), barium (Ba), and radium (Ra). The alkaline earth metals are shiny metals like those in Group ⅠA, but they are not as reactive.

The halogens are found on the right side of the periodic table in Group ⅦA. They include the elements fluorine (F), chlorine (Cl), bromine (Br), iodine (I), and astatine (At). The halogens, especially fluorine and chlorine, are highly reactive and form compounds with most of the elements.

The noble gases are found in Group ⅧA. They include helium (He), neon (Ne), argon (Ar), krypton (Kr), xenon (Xe), and radon (Rn). They are quite unreactive and are seldom found in combination with other elements.

Trends in Periodic Properties

The electron arrangements of atoms are an important factor in the physical and chemical properties of the elements. Now we will look at the valence electrons in atoms, the trends in atomic size, ionization energy, and metallic character. Known as periodic properties, there is a pattern of regular change going across a period, and then the trend is repeated again in each successive period.

Group Number and Valence Electrons

The chemical properties of representative elements in Groups ⅠA to ⅧA are mostly due to the valence electrons, which are the electrons in the outermost energy level. The group number gives the number of valence electrons for each group of representative elements. For example, all the elements in Group ⅠA have one valence electron. All the elements in Group ⅡA have two valence electrons. The halogens in Group ⅦA all have seven valence electrons.

Atomic Size

The size of an atom is determined by the distance of the valence electrons from the nucleus. For each group of representative elements, the atomic size *increases* going from the top to the bottom because the outermost electrons in each energy level are farther from the nucleus. For example, in Group ⅠA, Li has a valence electron in energy level 2; Na has a valence electron in energy level 3; and K has a valence electron in energy level 4. This means that a K atom is larger than a Na atom, and a Na atom is larger than a Li atom.

For the elements in a period, an increase in the number of protons in the nucleus increases the attraction for the outermost electrons. As a result, the outer electrons are pulled closer to the nucleus, which means that the size of representative elements *decreases* going from left to right across a period.

Vocabulary

organize ['ɔːgənaiz]　　*v.* 组织
type [taip]　　*n.* 类别
classification [ˌklæsifi'keiʃn]　　*n.* 分类
property ['prɔpəti]　　*n.* 性质；特性；性能
conductor [kən'dʌktə]　　*n.* 导体
relatively ['relətivli]　　*adv.* 相当地
vigorously ['vigərəsli]　　*adv.* 活泼地
melting points　熔点
hydrogen ['haidrədʒən]　　*n.* 氢
calcium ['kælsiəm]　　*n.* 钙
helium ['hiːliəm]　　*n.* 氦
magnesium [mæg'niːziəm]　　*n.* 镁
barium ['beəriəm]　　*n.* 钡
fluorine ['flɔːriːn]　　*n.* 氟
chlorine ['klɔːriːn]　　*n.* 氯
iodine ['aiədiːn]　　*n.* 碘
argon ['ɑːgɔn]　　*n.* 氩
beryllium [bə'riliəm]　　*n.* 铍
ionization [ˌaiənai'zeiʃn]　　*n.* 电离；离子化
metallic [mə'tælik]　　*adj.* 金属的
pattern ['pætn]　　*n.* 模式；方式；式样
valence ['væləns]　　*n.* 原子价
halogens ['hælədʒənz]　　*n.* 卤素类；卤素；卤素化合物；卤族
trends [trendz]　　*n.* 趋势；趋向；倾向；动态；动向
element ['elimənt]　　*n.* 元素，要素，原理，成分
period ['piəriəd]　　*n.* 周期

outermost ['autəməust]　*adj.* 最外层的；最外边的；最远的
determined [di'tə:mind]　*adj.* 决定
atomic [ə'tɔmik]　*adj.* 原子的
atom ['ætəm]　*n.* 原子
representative [,repri'zentətiv]　*adj.* 代表的
electron [ɪ'lektrɒn]　*n.* 电子
nucleus ['nju:kliəs]　*n.* 核
proton ['prəutɔn]　*n.* 质子

Lesson 3   The Nonmetal Elements

We noted earlier that nonmetals exhibit properties that are greatly different from those of the metals. As a rule, the nonmetals are poor conductors of electricity (graphitic carbon is an exception) and heat; they are brittle, are often intensely colored, and show an unusually wide range of melting and boiling points. Their molecular structures, usually involving ordinary covalent bonds, vary from the simple diatomic molecules of H_2, Cl_2, I_2, and N_2 to the giant molecules of diamond, silicon and boron.

The nonmetals that are gases at room temperature are the low-molecular weight diatomic molecules and the noble gases that exert very small intermolecular forces. As the molecular weight increases, we encounter a liquid (Br_2) and a solid (I_2) whose vapor pressures also indicate small intermolecular forces.

Simple diatomic molecules are not formed by the heavier members of Groups VA and VIA at ordinary conditions. This is in direct contrast to the first members of these groups, N_2 and O_2. The difference arises because of the lower stability of π bonds formed from p orbitals of the third and higher main energy levels as opposed to the second main energy level. The larger atomic radii and more dense electron clouds of elements of the third period and higher do not allow good parallel overlap of p orbitals necessary for a strong π bond. This is a general phenomenon—strong π bonds are formed only between elements of the second period. Thus, elemental nitrogen and oxygen form stable molecules with both σ and π bonds, but other members of their groups form more stable structures based on σ bonds only at ordinary conditions. Note that Group VIIA elements form diatomic molecules, but π bonds are not required for saturation of valence.

Sulfur exhibits allotropic forms. Solid sulfur exists in two crystalline forms and in an amorphous form. Rhombic sulfur is obtained by crystallization from a suitable solution, such as CS_2, and it melts at 112 °C. Monoclinic sulfur is formed by cooling melted sulfur and it melts at 119 °C. Both forms of crystalline sulfur melt into S-gamma, which is composed of S_8 molecules. The S_8 molecules are puckered rings and survive heating to about 160 °C. Above 160 °C, the S_8 rings break open, and some of these fragments combine with each other to form a highly viscous mixture of irregularly shaped coils. At a range of higher temperatures the liquid sulfur becomes so viscous that it will not pour from its container. The color also changes from straw yellow at sulfur's melting point to a deep reddish-brown as it becomes more viscous.

As the boiling point of 444 °C is approached, the large-coiled molecules of sulfur are partially degraded and the liquid sulfur decreases in viscosity. If the hot liquid sulfur is quenched by pouring it into cold water, the amorphous form of sulfur is produced. The structure of amorphous sulfur consists of large-coiled helices with eight sulfur atoms to each turn of the helix; the overall nature of amorphous sulfur is described as rubbery because it stretches much like ordinary rubber. In a few hours the amorphous sulfur reverts to small rhombic crystals and its rubbery property disappears.

Sulfur, an important raw material in industrial chemistry, occurs as the free element, as SO_2 in volcanic regions, as H_2S in mineral waters, and in a variety of sulfide ores such as iron pyrite FeS_2, zinc blende ZnS, galena PbS and such, and in common formations of gypsum $CaSO_4 \cdot 2H_2O$, anhydrite $CaSO_4$, and barites $BaSO_4 \cdot 2H_2O$. Sulfur, in one form or another, is used in large quantities for making sulfuric acid, fertilizers, insecticides, and paper.

Sulfur in the form of SO_2 obtained in the roasting of sulfide ores is recovered and converted to sulfuric acid, although in previous years much of this SO_2 was discarded through exceptionally tall smokestacks. Fortunately, it is now economically favorable to recover these gases, thus greatly reducing this type of atmospheric pollution. A typical roasting reaction involves the change:

$$2ZnS + 3O_2 \rightleftharpoons 2ZnO + 2SO_2$$

Phosphorus, below 800 °C consists of tetratomic molecules, P_4. Its molecular structure provides for a covalence of three, as may be expected from the three unpaired p electrons in its atomic structure, and each atom is attached to three others. Instead of a strictly orthogonal orientation, with the three bonds 90° to each other, the bond angles are only 60°. This supposedly strained structure is stabilized by the mutual interaction of the four atoms (each atom is bonded to the other three), but it is chemically the most active form of phosphorus. This form of phosphorus, the white modification, is spontaneously combustible in air. When heated to 260 °C it changes to red phosphorus, whose structure is obscure. Red phosphorus is stable in air but, like all forms of phosphorus, it should be handled carefully because of its tendency to migrate to the bones when ingested, resulting in serious physiological damage.

Carbon has the symbol C. However, its atoms can be arranged in different ways to give several different substances. Two forms of carbon—diamond and graphite—have been known since prehistoric times. In diamond, carbon atoms are arranged in a rigid structure. A diamond is transparent and harder than any other substance, whereas graphite is black and soft. In graphite, carbon atoms are arranged in sheets that slide over each other. Graphite is used as pencil lead, as lubricants, and as carbon fibers for the manufacture of lightweight golf clubs and tennis rackets. Two other forms of carbon have been discovered more recently. In the form called *Buckminsterfullerene* or *buckyball* (named after R. Buckminster Fuller, who popularized the geodesic dome), 60 carbon atoms are arranged as rings of five and six atoms to give a spherical, cage-like structure. When a fullerene structure is stretched out, it produces a cylinder with a diameter of only a few nanometers called a *nanotube*. Practical uses for buckyballs and nanotubes are not yet developed, but they are expected to find use in lightweight structural materials, heat conductors, computer parts, and medicine. Recent research has shown that carbon nanotubes (CNTs) can carry many drug molecules that can be released once the CNTs enter the targeted cells.

Vocabulary

diatomic molecules 双原子分子
electricity [ɪˌlek'trɪsəti] *n.* 电；电流；电力
silicon ['sɪlɪkən] *n.* 硅
molecular weight 分子量；分子质量
phenomenon [fə'nɒmɪnən] *n.* 现象
orbital ['ɔ:bɪt(ə)l] *n.* 轨道
crystalline ['krɪstəlaɪn] forms 结晶形态；晶型；晶型结构
crystallization 结晶；晶化；析晶
amorphous [ə'mɔ:fəs] form 非晶形成；无定形态；非晶态
melting point 熔点
straw yellow 淡黄色
sulfur ['sʌlfə(r)] *n.* 固体硫磺；固硫；固体硫
monoclinic [ˌmɒnəʊ'klɪnɪk] *adj.* 单斜晶系的；单斜的；单斜晶的
fragment ['frægmənt] *n.* 断片；残片
　　　　　　　　　　　v. 碎片化
boiling point 沸点；沸点温度
decrease 降低；减小；下降
stretch 伸长；伸张症；绵延
quenched [kwentʃt] *n.* 急冷；淬火的；消失过的
sulfide ores ['sʌlfaɪd] [ɔ:z] 硫化矿；含硫矿石；硫化物
zinc blende 闪锌矿；闪锌矿
pyrite ['paɪraɪt] *n.* 黄铁矿；天然的二硫化铁
roasting 焙烧；焙烤；烤烘用具

mutual interaction 双向互动；相互作用；相互影响
combustible [kəm'bʌstəbl] *adj.* 可燃性；易燃的；可燃烧的
phosphorus ['fɒsfərəs] *n.* 磷；磷素
orthogonal [ɔː'θɒgən(ə)l] *adj.* 矩形的；垂直的；正交性
carbon fibers 碳纤维；炭纤维；炭纤维的
transparent [træns'pærənt] *adj.* 透明；透明性
diameter [daɪ'æmɪtə(r)] *n.* 直径

Lesson 4 The Metallic Elements

An element that has metallic character is an element that loses valence electrons easily. Metallic character is more prevalent in the elements (metals) on the left side of the periodic table and decreases going from left to right across a period. The elements (nonmetals) on the right side of the periodic table do not easily lose electrons, which means they are the least metallic. Most of the metalloids between the metals and nonmetals tend to lose electrons, but not as easily as the metals. Thus, in Period 3, sodium, which loses electrons most easily, would be the most metallic. Going across from left to right in Period 3, metallic character decreases to argon, which has the least metallic character.

For elements in the same group of representative elements, metallic character increases going from top to bottom. Atoms at the bottom of any group have more electron levels, which makes it easier to lose electrons. Thus, the elements at the bottom of a group on the periodic table have lower ionization energy and are more metallic compared to the elements at the top.

Elements in Group ⅠB and ⅡB have a greater bulk use as metals than in compounds, and their physical properties vary widely.

Gold is the most malleable and ductile of the metals. It can be hammered into sheets of 0.00001 inch in thickness; one gram of the metal can be drawn into a wire 1.8 mi in length. Copper and silver are also metals that are mechanically easy to work. Zinc is a little brittle at ordinary temperatures, but may be rolled into sheets at between 120 °C to 150 °C; it becomes brittle again about 200 °C. The low-melting temperatures of zinc contribute to the preparation of zinc-coated iron. galvanized iron; clean iron sheet may be dipped into vats of liquid zinc in its preparation. A different procedure is to sprinkle or air blast zinc dust onto hot iron sheeting for a zinc melt and then coating.

Cadmium has specific uses because of its low-melting temperature in a number of alloys. Cadmium rods are used in nuclear reactors because the metal is a good neutron absorber.

Mercury vapor and its salts are poisonous, though the free metal may be taken internally under certain conditions. Because of its relatively low boiling point and hence volatile nature, free mercury should never be allowed to stand in an open container in the

Part 1 Elements and Inorganic Compounds

laboratory. Evidence shows that inhalation of its vapors is injurious. Blood, urine, and hair samples are used to test for mercury.

The metal alloys readily with most of the metals (except iron and platinum) to form amalgams, the name given to any alloy of mercury.

Compounds of Copper

Copper sulfate, or blue vitriol ($CuSO_4 \cdot 5H_2O$) is the most important and widely used salt of copper. On heating, the salt slowly loses water to form first the trihydrate ($CuSO_4 \cdot 3H_2O$), then the monohydrate ($CuSO_4 \cdot H_2O$), and finally the white anhydrous salt. The anhydrous salt is often used to test for the presence of water in organic liquids. For example, some of the anhydrous copper salt added to alcohol (which contains water) will turn blue because of the hydration of the salt.

Copper sulfate is used in electroplating. Fishermen dip their nets in copper sulfate solution to inhibit the growth of organisms that would rot the fabric. Paints specifically formulated for use on the bottoms of marine craft contain copper compounds to inhibit the growth of barnacles and other organisms.

When dilute ammonium hydroxide is added to a solution of copper (Ⅰ) ions, a greenish precipitate of $Cu(OH)_2$ or a basic copper (Ⅰ) salt is formed. This dissolves as more ammonium hydroxide is added. The excess ammonia forms an ammoniated complex with the copper (Ⅰ) ion of the composition, $[Cu(NH_3)_4]^{2+}$. This ion is only slightly dissociated; hence in an ammoniacal solution very few copper (Ⅰ) ions are present. Insoluble copper compounds, execpt copper sulfide, are dissolved by ammonium hydroxids. The formation of the copper (Ⅰ) ammonia ion is often used as a test for Cu^{2+} because of its deep, intense blue color.

Copper (Ⅰ) ferrocyanide $[Cu_2Fe(CN)_6]$ is obtained as a reddish-brown precipitate on the addition of a soluble ferrocyanide to a solution of copper (Ⅰ)ions. The formation of this salt is also used as a test for the presence of copper (Ⅰ) ions.

Compounds of Silver and Gold

Silver nitrate, sometimes called lunar caustic, is the most important salt of silver. It melts readily and may be cast into sticks for use in cauterizing wounds. The salt is prepared by dissolving silver in nitric acid and evaporating the solution. The salt is the starting material for most of the compounds of silver, including the halides used in photography. It is readily reduced by organic reducing agents, with the formation of a black deposit of finely divided silver; this action is responsible for black spots left on the fingers from the handling of the salt.

Indelible marking inks and pencils take advantage of this property of silver nitrate. The halides of silver, except the fluoride, are very insoluble compounds and may be precipitated by the addition of a solution of silver salt to a solution containing chloride, bromide, or iodide ions.

The addition of a strong base to a solution of a silver salt precipitates brown silver oxide (Ag_2O). One might expect the hydroxide of silver to precipitate, but it seems likely that silver hydroxide is very unstable and breaks down into the oxide and water—if, indeed, it is ever formed at all. However, since a solution of silver oxide is definitely basic, there must be hydroxide ions present in solution.

$$Ag_2O + H_2O \rightleftharpoons 2Ag^+ + 2OH^-$$

Because of its inactivity, gold forms relatively few compounds. Two series of compounds are known—monovalent and trivalent. Monovalent (aurous) compounds resemble silver compounds (aurous chloride is water insoluble and light sensitive), while the higher valence (auric) compounds tend to form complexes. Gold is resistant to the action of most chemicals—air, oxygen, and water have no effect.

The common acids do not attack the metal, but a mixture of hydrochloric and nitric acids (aqua regia) dissolves it to form gold (I) chloride or chloroauric acid. The action is probably due to free chlorine present in the aqua regia.

$$3HCl + HNO_3 \longrightarrow NOCl + Cl_2 + 2H_2O$$

$$2Au + 3Cl_2 \longrightarrow 2AuCl_3$$

$$AuCl_3 + HCl \longrightarrow HAuCl_4$$

chloroauric acid ($HAuCl_4 \cdot H_2O$ crystallizes from solution).

Compounds of Zinc

Zinc is fairly high in the activity series. It reacts readily with acids to produce hydrogen and displaces less active metals from their salts. The action of acids on impure zinc is much more rapid than on pure zinc, since bubbles of hydrogen gas collect on the surface of pure zinc and slow down the action. If another metal is present as an impurity, the hydrogen is liberated from the surface of the contaminating metal rather than from the zinc. An electric couple to facilitate the action is probably set up between the two metals.

$$Zn + 2H^+ \longrightarrow Zn^{2+} + H_2$$

Zinc oxide (ZnO), the most widely used zinc compound, is a white powder at ordinary temperatures, but changes to yellow on heating. When cooled, it again becomes white. Zinc oxide is obtained by burning zinc in air, by heating the basic carbonate, or by roasting the sulfide. The principal use of ZnO is as a filler in rubber manufacture, particularly in automobile tires. As a body for paints it has the advantage over white lead of not darkening on exposure to an atmosphere containing hydrogen sulfide. Its covering power, however, is inferior to that of white lead.

Vocabulary

prevalent ['prevələnt]　*adj.* 流行的；盛行的；普遍存在的
periodic [ˌpɪəri'ɒdɪk]　*adj.* 周期性；周期的
metalloids ['metlɔɪd]　*n.* 准金属
　　　　　　　　　　adj. 似金属的
argon ['ɑːɡɒn]　*n.* 氩
ionization [ˌaɪənaɪ'zeɪʃn]　*n.* 电离；离子化
bulk [bʌlk]　*n.* 主体；大部分；（大）体积
malleable ['mæliəbl]　*adj.* 有延展性的；可塑的；可锻造的
ductile ['dʌktaɪl]　*adj.* 可拉成细丝的；可延展的；有延性的
hammered ['hæməd]　*n.* 锤子；榔头
　　　　　　　　　　v.（用锤子）敲，锤打；反复敲打
sheets [ʃiːts]　*n.* 薄片
　　　　　　　　adj. 片状的
brittle ['brɪtl]　*adj.* 易碎的；脆性的；脆弱的
coated ['kəʊtɪd]　*n.* 涂料层
　　　　　　　　　vt. 给……涂上一层；用……覆盖
galvanized ['ɡælvənaɪzd]　*vt.* 镀锌；激励；电镀
sprinkle ['sprɪŋkl]　*v.* 撒；洒；把……撒
blast [blɑːst]　*n.* 爆炸；把……炸成碎片
nuclear ['njuːkliə(r)]　*adj.* 原子能的；核能的
neutron ['njuːtrɒn]　*n.* 中子
mercury vapor ['mɜːkjəri] ['veɪpə(r)]　*n.* 水银蒸气
internally [ɪn'tɜːnəli]　*adv.* 内部地；内在地
volatile ['vɒlətaɪl]　*adj.* 不稳定的；易挥发的；易变的
inhalation [ˌɪnhə'leɪʃn]　*n.*（空气等的）吸入；吸入剂；吸入药
injurious [ɪn'dʒʊəriəs]　*adj.* 造成伤害的；有害的
urine ['jʊərɪn]　*n.* 尿
platinum ['plætɪnəm]　*n.* 铂；白金
amalgams [ə'mælɡəmz]　*n.* 汞合金；混合物

vitriol ['vɪtrɪəl]　　n. 硫酸盐，刻薄话；蓝矾
trihydrate [traɪ'haɪdreɪt]　n. 三水合物
anhydrous [æn'haɪdrəs]　adj. 无水的（尤指结晶水）
alcohol ['ælkəhɒl]　n. 酒精；酒；醇
hydration [haɪ'dreɪʃən]　n. 水化（合）作用；水化
electroplating [ɪ'lektrəpleɪtɪŋ]　v. 电镀
rot [rɒt]　v. 腐烂；（使）腐败变质
fabric ['fæbrɪk]　n. 织物；结构
craft [krɑːft]　n. 船；飞行器
barnacles ['bɑːnəklz]　n. 藤壶
dilute [daɪ'luːt]　v. （使）稀释；削弱，降低
ammonium [ə'məʊnɪəm]　n. 铵
greenish ['griːnɪʃ]　adj. 呈绿色的，微绿的
excess [ɪk'ses;'ekses]　n. 过量；无节制
ferrocyanide [ferəʊ'saɪənaɪd]　n. 氰亚铁酸盐，亚铁氰化物
nitrate ['naitreit]　n. 硝酸盐
lunar caustic ['luːnə(r)]['kɔːstɪk]　王水
cast [kɑːst]　v. 浇铸
cauterizing ['kɔːtəraɪzɪŋ]　v. 灼，腐蚀
nitric ['naitrik]　acid. 硝酸
evaporating [ɪ'væpəreɪtɪŋ]　v. 蒸发
halide ['hælaid]　n. 卤化物
reduced [rɪ'djuːst]　vt. 还原，减小
deposit [dɪ'pɒzɪt]　n. 沉淀
finely ['faɪnli]　adv. 细小地
silver nitrate /'sɪlvə(r)/英式　['naitreit]　n. 硝酸银
halide ['hæˌlaɪd]　n. 卤化物
fluoride ['flɔːraɪd]　n. 氟化物
precipitated [prɪ'sɪpɪteɪtɪd]　n. 沉淀物；析出物
bromide ['brəʊmaɪd]　n. 溴化物
iodide ['aɪədaɪd]　n. 碘化物，碘负离子
base [beis]　n. 碱
silver ['sɪlvə(r)]　n. 银
hydroxide [haɪ'drɒksaɪd]　n. 氢氧化物
precipitate [prɪ'sɪpɪteɪt]　n. 沉淀物
oxide ['ɒksaɪd]　n. 氧化物

Part 1　Elements and Inorganic Compounds

definitely ['defɪnətli]　*adv.* 明确地，肯定地
monovalent [ˌmɒnə(ʊ)'veɪl(ə)nt]　*adj.* 单价的；一价的
aurous ['ɔːrəs]　*adj.* 含金的；亚金的
chloride ['klɔːraɪd]　*n.* 氯化物
valence ['væləns]　*n.* 价；原子价
hydrochloric ['haɪdrə'klɒrɪk]　*adj.* 氯化氢的，盐酸的
nitric ['naɪtrɪk]　acids *n.* 硝酸
aqua regia [ˌækwə'riːdʒə]　*n.* 王水
chloride ['klɔːraɪd]　*n.* 氯化物
chloroauric ['klɔːrə'ɔːrɪk] acid　*n.* 氯金酸
chlorine ['klɔːriːn]　*n.* 氯气
crystallizes ['krɪstəlaɪzɪz]　*v.* 使结晶
solution [sə'luːʃn]　*n.* 解决方案；溶液
impurity /ɪm'pjʊərəti/ 英式　*n.* 杂质；不纯
liberated ['lɪbəreɪtɪd]　*v.* 解放；释放
contaminating [kən'tæmɪneɪtɪŋ]　*v.* 污染
facilitate [fə'sɪlɪteɪt]　*v.* 促进
filler ['fɪlə]　*n.* 填充物
powder ['paʊdə(r)]　*n.* 粉末
obtained [əb'teɪnd]　*v.*（尤指通过艰难的过程）得到，获得
basic ['beɪsɪk]　*adj.* 碱式的
carbonate ['kɑːbənət]　*n.* 碳酸盐
roasting ['rəʊstɪŋ]　*v.* 煅烧
sulfide ['sʌlfaɪd]　*n.* 硫化物
rubber ['rʌbə]　*n.* [化] 橡胶
tires [taɪəz]　*n.* 轮胎
white lead [ˌwaɪt 'led]　*n.* 白铅
atmosphere ['ætməsfɪə]　*n.* 大气，空气
volcanic [vɒl'kænɪk]　*adj.* 火山的，火山引发的
region ['riːdʒən]　*n.* 地区；地域；领域
pyrite ['paɪraɪt]　*n.* 黄铁矿
blende [blend]　*n.* 闪锌矿；褐色闪光矿物
galena [gə'liːnə]　*n.* 方铅矿；硫化铅
gypsum ['dʒɪpsəm]　*n.* 石膏，石膏肥料
anhydrite [æn'haɪdraɪt]　*n.* 硬石膏；无水石膏
barite ['beəraɪt]　*n.* 重晶石

- 19 -

fertilizer ['fɜ:təlaɪzə]　　*n.* 肥料；化肥
insecticide [ɪn'sektɪsaɪd]　　*n.* 杀虫剂
diamond ['daɪəmənd]　　*n.* 钻石；金刚钻
graphite ['græfaɪt]　　*n.* 石墨
prehistoric [ˌpri:hɪ'stɒrɪk]　　*adj.* 有文字记载历史以前的；史前的
lubricants ['lu:brɪkənts]　　*n.* 润滑剂；润滑油
fibers [faɪbəz]　　*n.* 纤维；织物；纤维素
manufacture [ˌmænju'fæktʃə]　　生产；制造；产品
geodesic [ˌdʒi:əʊ'desɪk]　　*adj.* 网格状的；网架状的
spherical ['sferɪkl]　　*adj.* 球形的；球体的
fullerene ['fʊləri:n]　　*n.* 富勒烯
cylinder ['sɪlɪndə]　　*n.* 圆柱体；圆柱形物
diameter [daɪ'æmɪtə]　　*n.* 直径；直径长；横度
nanometer ['nænəʊmi:tə(r)]　　*n.* 纳米
nanotube ['nænəʊtju:b]　　*n.* 纳米管

Part 1   Elements and Inorganic Compounds

Lesson 5   Ionic and Molecular Compounds

In nature, atoms of almost all the elements on the periodic table are found in combination with other atoms. Only the atoms of the noble gases (He, Ne, Ar, Kr, Xe, and Rn) do not combine in nature with other atoms. A compound is composed of two or more elements, with a definite composition. Compounds may be ionic or molecular.

In an ionic compound, one or more electrons are transferred from metals to nonmetals, which form positive and negative ions. The attraction between these ions is called an ionic bond. We utilize ionic compounds every day, including salt, NaCl, and baking soda, $NaHCO_3$. Milk of magnesia, $Mg(OH)_2$, or calcium carbonate, $CaCO_3$, may be taken to settle an upset stomach. In a mineral supplement, iron may be present as iron (Ⅱ) sulfate, $FeSO_4$, iodine as potassium iodide, KI, and manganese as manganese (Ⅱ) sulfate, $MnSO_4$. Some sunscreens contain zinc oxide, ZnO, and tin (Ⅱ) fluoride, SnF_2, in toothpaste provides fluoride to help prevent tooth decay. Precious and semiprecious gemstones are examples of ionic compounds called minerals that are cut and polished to make jewelry. For example, sapphires and rubies are made of aluminum oxide, Al_2O_3. Impurities of chromium make rubies red, and iron and titanium make sapphires blue.

A molecular compound consists of two or more nonmetals that share one or more valence electrons. The resulting molecules are held together by covalent bonds. There are many more molecular compounds than there are ionic ones. For example, water (H_2O) and carbon dioxide (CO_2) are both molecular compounds. Molecular compounds consist of molecules, which are discrete groups of atoms in a definite proportion. A molecule of water (H_2O) consists of two atoms of hydrogen and one atom of oxygen. When you have iced tea, perhaps you add molecules of sugar ($C_{12}H_{22}O_{11}$), which is a molecular compound. Other molecular compounds include propane (C_3H_8), alcohol (C_2H_6O), the antibiotic amoxicillin ($C_{16}H_{19}N_3O_5S$), and the antidepressant Prozac ($C_{17}H_{18}F_3NO$).

Ionic compounds

Ionic compounds consist of positive and negative ions. The ions are held together by strong electrical attractions between the oppositely charged ions, called ionic bonds. The positive ions are formed by metals losing electrons, and the negative ions are formed when nonmetals gain electrons. Even though noble gases are nonmetals, they already have stable electron configurations and do not form compounds

Properties of Ionic Compounds

The physical and chemical properties of an ionic compound such as NaCl are very different from those of the original elements. For example, the original elements of NaCl were sodium, which is a soft, shiny metal, and chlorine, which is a yellow-green poisonous gas. However, when they react and form positive and negative ions, they produce NaCl, which is ordinary table salt, a hard, white, crystalline substance that is important in our diet.

In a crystal of NaCl, the larger Cl^- ions are arranged in a three-dimensional structure in which the smaller Na^+ ions occupy the spaces between the Cl^- ions. In this crystal, every Na^+ ion is surrounded by six Cl^- ions, and every Cl^- ion is surrounded by six Na^+ ions. Thus, there are many strong attractions between the positive and negative ions, which account for the high melting points of ionic compounds. For example, the melting point of NaCl is 801 °C. At room temperature, ionic compounds are solids.

Chemical Formulas of Ionic Compounds

The chemical formula of a compound represents the symbols and subscripts in the lowest whole-number ratio of the atoms or ions. In the formula of an ionic compound, the sum of the ionic charges in the formula is always zero. *Thus, the total amount of positive charge is equal to the total amount of negative charge.* For example, to achieve a stable electron configuration, a Na atom (metal) loses its one valence electron to form Na^+, and one Cl atom (nonmetal) gains one electron to form a Cl^- ion. The formula NaCl indicates that the compound has charge balance because there is one sodium ion, Na^+, for every chloride ion, Cl^-. Although the ions are positively or negatively charged, they are not shown in the formula of the compound.

Naming Ionic Compounds

In the name of an ionic compound made up of two elements, the name of the metal ion, which is written first, is the same as its element name. The name of the nonmetal ion is obtained by using the first syllable of its element name followed by *ide*. In the name of any ionic compound, a space separates the name of the cation from the name of the anion. Subscripts are not used; they are understood because of the charge balance of the ions in the compound (see Table 5.1).

Table 5.1 Names of Some Ionic Compounds

Compound	Metal	Ion	Nonmetal
KI	K^+	I^-	
	Potassium	Iodide	Potassium iodide
$MgBr_2$	Mg^{2+}	Br^-	
	Magnesium	Bromide	Magnesium bromide
Al_2O_3	Al^{3+}	O^{2-}	
	Aluminum	Oxide	Aluminum oxide

Polyatomic Ions

An ionic compound may also contain a *polyatomic ion* as one of its cations or anions. A polyatomic ion is a group of covalently bonded atoms that has an overall ionic charge. Most polyatomic ions consist of a nonmetal such as phosphorus, sulfur, carbon, or nitrogen covalently bonded to oxygen atoms. Almost all the polyatomic ions are anions with charges of 1-, 2-, or 3-. Only one common polyatomic ion, NH_4^+, has a positive charge.

Names of Polyatomic Ions

The names of the most common polyatomic ions end in *ate*, such as nitrate and sulfate. When a related ion has one less oxygen atom, the *ite* ending is used for its name such asnitrite and sulfite. Recognizing these endings will help you identify polyatomic ions in the name of a compound. The hydroxide ion (OH^-) and cyanide ion (CN^-) are exceptions to this naming pattern.

By learning the formulas, charges, and names of the polyatomic ions shown in bold type in Table 5.2, you can derive the related ions. Note that both the *ate* ion and *ite* ion of a particular nonmetal have the same ionic charge. For example, the sulfate ion is SO_4^{2-}, and the sulfite ion, which has one less oxygen atom, is SO_3^{2-}. Phosphate and phosphite ions each have a 3-charge; nitrate and nitrite each have a 1-charge; and perchlorate, chlorate, chlorite, and hypochlorite all have a 1-charge. The halogens form four different polyatomic ions with oxygen. The formula of hydrogen carbonate, or *bicarbonate*, is written by placing a hydrogen in front of the polyatomic ion formula for carbonate (CO_3^{2-}), and the charge is decreased from 2- to 1- to give HCO_3^-.

$$H^+ + CO_3^{2-} \rightleftharpoons HCO_3^-$$

Table 5.2 Names and Formulas of Some Common Polyatomic Ions

Nonmetal	Formula of Ion	Name of Ion
Hydrogen	OH^-	Hydroxide
Nitrogen	NH_4^+	Ammonium
	NO_3^-	Nitrate
	NO_2^-	Nitrite
Chlorine	ClO_4^-	Perchlorate
	ClO_3^-	Chlorate
	ClO_2^-	Chlorite
	ClO^-	Hypochlorite
Carbon	CO_3^{2-}	Carbonate
	HCO_3^-	Hydrogen carbonate (or bicarbonate)
	CN^-	Cyanide
	$C_2H_3O_2^-$	Acetate
Sulfur	SO_4^{2-}	Sulfate
	HSO_4^-	Hydrogen sulfate (or bisulfate)
	SO_3^{2-}	Sulfite
	HSO_3^-	Hydrogen sulfite (or bisulfite)
Phosphorus	PO_4^{3-}	Phosphate
	HPO_4^{2-}	Hydrogen phosphate
	$H_2PO_4^-$	Dihydrogen phosphate
	PO_3^{3-}	Phosphite

Molecular Compounds

A molecular compound contains two or more nonmetals that form *covalent bonds*. Because nonmetals have high ionization energies, valence electrons are shared by nonmetal atoms to achieve stability. When atoms share electrons, the bond is a covalent bond. When two or more atoms share electrons, they form a molecule.

Names and Formulas of Molecular Compounds

When naming a molecular compound, the first nonmetal in the formula is named by itselement name; the second nonmetal is named using the first syllable of its element name, followed by *ide*. When a subscript indicates two or more atoms of an element, a prefix is shown in front of its name. Table 5.3 lists prefixes used in naming molecular compounds. The names of molecular compounds need prefixes because several different compounds can

Part 1 Elements and Inorganic Compounds

be formed from the same two nonmetals. For example, carbon and oxygen can form two different compounds, carbon monoxide, CO, and carbon dioxide, CO_2, in which the number of atoms of oxygen in each compound is indicated by the prefixes *mono* or *di* in their names. When the vowels *o* and *o* or *a* and *o* appear together, the first vowel is omitted, as in carbon monoxide. In the name of a molecular compound, the prefix *mono* is usually omitted, as in NO, nitrogen oxide. Traditionally, however, CO is named carbon monoxide. Table 5.4 lists the formulas, names, and commercial uses of some molecular compounds.

Table 5.3 Prefixes Used in Naming Molecular Compounds

1 mono	6 hexa
2 di	7 hepta
3 tri	8 octa
4 tetra	9 nona
5 penta	10 deca

Table 5.4 Some Common Molecular Compounds

Formula	Name	Commercial Uses
CS_2	Carbon disulfide	Manufacture of rayon
CO_2	Carbon dioxide	Fire extinguishers, dry ice, propellant in aerosols, carbonation of beverages
NO	Nitrogen oxide	Stabilizer
N_2O	Dinitrogen oxide	Inhalation anesthetic, "laughing gas"
SO_2	Sulfur dioxide	Preserving fruits, vegetables; disinfectant in breweries; bleaching textiles
SO_3	Sulfur trioxide	Manufacture of explosives
SF_6	Sulfur hexafluoride	Electrical circuits

Summary of Naming Ionic and Molecular Compounds

We have now examined strategies for naming ionic and molecular compounds. In general, compounds having two elements are named by stating the first element name, followed by the name of the second element with an *ide* ending. If the first element is a metal, the compounds usually ionic; if the first element is a nonmetal, the compound is usually molecular. For ionic compounds, it is necessary to determine whether the metal can form more than one type of positive ion; if so, a Roman numeral following the name of the metal indicates the particular ionic charge. One exception is the ammonium ion, NH_4^+, which is also written first as a positively charged polyatomic ion. Ionic compounds having three or more elements include some type of polyatomic ion. They are named by ionic rules but have an *ate* or *ite* ending when the polyatomic ion has a negative charge. In naming molecular compounds having two elements, prefixes are necessary to indicate two or more atoms of each nonmetal.

Vocabulary

molecular [mə'lekjələ(r)] *adj.* 分子的；由分子组成的

utilize ['ju:təlaɪz] *v.* 利用

magnesia [mæg'ni:ʒə; mæg'ni:zɪə] *n.* 氧化镁；苦土；水合碳酸镁

supplement ['sʌplɪmənt] *v.* 增补（物）；补品

sulfate ['sʌlfeɪt] *n.* [无化] 硫酸盐

iodine ['aɪədi:n] *n.* 碘；碘酒

toothpaste ['tu:θpeɪst] *n.* 牙膏

fluoride ['flɔ:raɪd;'flʊəraɪd] *n.* 氟化物

semiprecious [ˌsemɪ'preʃəs] *adj.* 次珍贵的；准宝石的

impurities [ɪm'pjʊərɪtɪz] *n.* 杂质（impurity 的复数）

subscripts ['sʌbˌskrɪpts] *n.* 下标

dimensional [daɪ'menʃənl] *adj.* 维的；尺寸的；空间的

poisonous ['pɔɪzənəs] *adj.* 有毒的；令人厌恶的；恶毒的

oppositely ['ɒpəzɪtlɪ] *adv.* 相对地；反向地；在相反的位置；面对面

react [ri'ækt] *v.* 作出反应；反应；反对

crystalline ['krɪstəlaɪn] *adj.* 结晶的；晶状的

represents [ˌreprɪ'zents] *vt.* 表现；代表

valence ['væləns; 'veɪləns] *n.* 化合价

configuration [kənˌfɪɡə'reɪʃn] *n.* 构造；结构；配置

syllable ['sɪləbl] *n.* 音节

subscript ['sʌbskrɪpt] *n.* 下标，脚注

polyatomic [ˌpɒlɪə'tɒmɪk] *adj.* 多原子的

phosphorus ['fɒsfərəs] *n.* 磷；磷光体

covalence [kəʊ'veɪləns] *n.* 共价

hydroxide [haɪ'drɒksaɪd] *n.* 氢氧化物，羟化物

cyanide ['saɪənaɪd] *n.* 氰化物

perchlorate [pə'klɔ:reɪt] *n.* 高氯酸盐

formula ['fɔ:mjələ] *n.* 公式，方程式

bicarbonate [ˌbaɪˈkɑːbənət]　　*n.* 重碳酸盐
subscript [ˈsʌbskrɪpt]　　*adj.* 下标的
omit [əˈmɪt]　　*v.* 忽略；漏掉；遗漏；省略
ionization /ˌaɪənaɪˈzeɪʃn/英式　　*n.* 电离；离子化
commercial /kəˈmɜːʃl/英式　　*adj.* 贸易的；商业的；赢利的
vowel [ˈvaʊəl]　　*n.* 元音；元音字母

Lesson 6 Acids, Bases and Equilibrium

Acids and bases are important substances in health, industry, and the environment. One of the most common characteristics of acids is their sour taste. Lemons and grapefruits are sour because they contain organic acids such as citric and ascorbic acid (vitamin C). Vinegar tastes sour because it contains acetic acid. We produce lactic acid in our muscles when we exercise. Acid from bacteria turns milk sour in the production of yogurt or cottage cheese. We have hydrochloric acid in our stomachs that helps us digest food. Sometimes we take antacids, which are bases such as sodium bicarbonate or milk of magnesia, to neutralize the effects of too much stomach acid.

In the environment, the acidity, or pH, of rain, water, and soil can have significant effects. When rain becomes too acidic, it can dissolve marble statues and accelerate the corrosion of metals. In lakes and ponds, the acidity of water can affect the ability of plants and fish to survive. The acidity of soil around plants affects their growth. If the soil pH is too acidic or too basic, the roots of the plant cannot take up some nutrients. Most plants thrive in soil with a nearly neutral pH, although certain plants such as orchids, camellias, and blueberries require a more acidic soil. Major changes in the pH of the body fluids can severely affect biological activities within the cells. Buffers are present to prevent large fluctuations in pH.

Acids and Bases

The term *acid* comes from the Latin word *acidus*, which means "sour". We are familiar with the sour tastes of vinegar and lemons and other common acids in foods. In 1887, the Swedish chemist Svante Arrhenius was the first to describe acids as substances that produce hydrogen ions (H^+) when they dissolve in water. Because acids produce ions in water, they are also electrolytes. For example, hydrogen chloride ionizes completely in water to give hydrogen ions, H^+, and chloride ions, Cl^-. It is the hydrogen ions that give acids a sour taste, change blue litmus indicator to red, and corrode some metals.

Part 1   Elements and Inorganic Compounds

Naming Acids

Acids dissolve in water to produce hydrogen ions, along with a negative ion that may be a simple nonmetal anion or a polyatomic ion. When an acid dissolves in water to produce a hydrogen ion and a simple nonmetal anion, the prefix *hydro* is used before the name of the nonmetal, and its *ide* ending is changed to *ic acid*. For example, hydrogen chloride (HCl) dissolves in water to form HCl (aq), which is named hydrochloric acid. An exception is hydrogen cyanide (HCN), which dissolves in water to form hydrocyanic acid, HCN (aq). When an acid contains oxygen, it dissolves in water to produce a hydrogen ion and an oxygen-containing polyatomic anion. The most common form of an oxygen-containing acid has a name that ends with *ic acid*. The name of its polyatomic anion ends in *ate*. An acid that contains one less oxygen atom than the common form is named as an *ous acid*. The name of its polyatomic anion ends with *ite* (see Table 6.1). By learning the names of the most common acids, we can derive the names of the corresponding *ous acids* and their polyatomic anions.

Table 6.1   Names of Common Acids and Their Anions

Acid	Name of Acid	Anion	Name of Anion
HCl	Hydrochloric acid	Cl^-	Chloride
HBr	Hydrobromic acid	Br^-	Bromide
HI	Hydroiodic acid	I^-	Iodide
HCN	Hydrocyanic acid	CN^-	Cyanide
HNO_3	Nitric acid	NO_3^-	Nitrate
HNO_2	Nitrous acid	NO_2^-	Nitrite
H_2SO_4	Sulfuric acid	SO_4^{2-}	Sulfate
H_2SO_3	Sulfurous acid	SO_3^{2-}	Sulfite
H_2CO_3	Carbonic acid	CO_3^{2-}	Carbonate
$HC_2H_3O_2$	Acetic acid	$C_2H_3O_2^-$	Acetate
H_3PO_4	Phosphoric acid	PO_4^{3-}	Phosphate
H_3PO_3	Phosphorous acid	PO_3^{3-}	Phosphite
$HClO_3$	Chloric acid	ClO_3^-	Chlorate
$HClO_2$	Chlorous acid	ClO_2^-	Chlorite

Bases

You may be familiar with some household bases such as antacids, drain openers, and oven cleaners. According to the Arrhenius theory, bases are ionic compounds that dissociate into a metal ion and hydroxide ions (OH^-) when they dissolve in water. Thus, Arrhenius bases are also electrolytes. For example, sodium hydroxide is an Arrhenius base that dissociates in water to give sodium ions, Na^+, and hydroxide ions, OH^-. Most Arrhenius bases are formed from Groups ⅠA (1) and ⅡA (2) metals, such as NaOH, KOH, LiOH, and $Ca(OH)_2$. The hydroxide ions (OH^-) give Arrhenius bases common characteristics such as a bitter taste and soapy, slippery feel. A base turns litmus indicator blue and phenolphthalein indicator pink.

Naming Bases

Typical Arrhenius bases are named as hydroxides (see Table 6.2).

Table 6.2　Names for Some Bases

Base	Name
LiOH	Lithium hydroxide
NaOH	Sodium hydroxide
KOH	Potassium hydroxide
$Ca(OH)_2$	Calcium hydroxide
$Al(OH)_3$	Aluminum hydroxide

Brønsted-Lowry Acids and Bases

In 1923, J. N. Brønsted in Denmark and T. M. Lowry in Great Britain expanded the definition of acids and bases to include bases that do not contain OH^- ions. A Brønsted-Lowry acid can donate a hydrogen ion (H^+) to another substance, and a Brønsted-Lowry base can accept a hydrogen ion (H^+). A Brønsted-Lowry acid is a substance that donates H^+. A Brønsted-Lowry base is a substance that accepts H^+. A free hydrogen ion, H^+, does not actually exist in water. Its attraction to polar water molecules is so strong that the H^+ bonds to a water molecule and forms a hydronium ion, H_3O^+.

$$H_2O + H^+ \longrightarrow H_3O^+$$

We can write the formation of a hydrochloric acid solution as a transfer of H^+ from hydrogen chloride to water. By accepting H^+ in the reaction, water is acting as a base according to the Brønsted-Lowry concept.

$$HCl + H_2O \longrightarrow H_3O^+ + Cl^-$$

Ammonia, NH_3, acts as a base by accepting H^+ when it reacts with water. Because the nitrogen atom of NH_3 has a stronger attraction for H^+ than the oxygen of water, water acts as an acid by donating H^+.

$$NH_3 + H_2O \longrightarrow NH_4^+ + OH^-$$

Part 1   Elements and Inorganic Compounds

Strengths of Acids and Bases

The strength of an acid is determined by the moles of H_3O^+ that are produced for each mole of acid that dissolves. The strength of a base is determined by the moles of OH^- that are produced for each mole of base that dissolves. Strong acids and strong bases ionize completely in water whereas weak acids and weak bases ionize only slightly in water, leaving most of the initial acid or base as molecules.

Strong and Weak Acids

Strong acids are examples of strong electrolytes because they donate H^+ so easily that their ionization in water is virtually complete. For example, when HCl, a strong acid, ionizes in water, H^+ is transferred to H_2O; the resulting solution contains only the ions H_3O^+ and Cl^-. We consider the reaction of HCl in H_2O as going 100% to products. Thus, the equation for a strong acid such as HCl is written with a single arrow to the products.

$$HCl(g) + H_2O(l) \longrightarrow H_3O^+(aq) + Cl^-(aq)$$

There are only six common strong acids. All other acids are weak (see Table 6.3).

Weak acids are weak electrolytes because they ionize slightly in water, which produces only a few ions. A solution of a strong acid contains all ions, whereas a solution of a weak acid contains mostly molecules and few ions.

Table 6.3   Common Strong and Weak Acids

Strong Acids	
Hydroiodic acid	HI
Hydrobromic acid	HBr
Perchloric acid	$HClO_4$
Hydrochloric acid	HCl
Sulfuric acid	H_2SO_4
Nitric acid	HNO_3
Weak Acids	
Hydronium ion	H_3O^+
Hydrogen sulfate ion	HSO_4^-
Phosphoric acid	H_3PO_4
Hydrofluoric acid	HF
Nitrous acid	HNO_2
Acetic acid	$HC_2H_3O_2$

Continued Table

Carbonic acid	H_2CO_3
Hydrosulfuric acid	H_2S
Dihydrogen phosphate	$H_2PO_4^-$
Ammonium ion	NH_4^+
Hydrocyanic acid	HCN
Bicarbonate ion	HCO_3^-
Hydrogen sulfide ion	HS^-
Water	H_2O

Strong and Weak Bases

As strong electrolytes, strong bases dissociate completely in water. Because these strong bases are ionic compounds, they dissociate in water to give an aqueous solution of metal ions and hydroxide ions (see Table 6.4). The Group ⅠA (1) hydroxides are very soluble in water, which can give high concentrations of OH^- ions. For example, when KOH forms a KOH solution, it contains only the ions K^+ and OH^-.

$$KOH(s) \xrightarrow{H_2O} K^+(aq) + OH^-(aq)$$

A few strong bases are less soluble in water, but what does dissolve dissociates completely as ions.

Table 6.4　Common Strong Bases

Strong Bases	
Lithium hydroxide	LiOH
Sodium hydroxide	NaOH
Potassium hydroxide	KOH
Strontium hydroxide	$Sr(OH)_2$
Calcium hydroxide	$Ca(OH)_2$
Barium hydroxide	$Ba(OH)_2$

Acid-Base Equilibrium

As we have seen, reactants in acid-base reactions are not always completely converted to products because a reverse reaction takes place in which products form reactants. A reversible reaction proceeds in both the forward and reverse directions. That means there are two reactions taking place: One is the reaction in the forward direction, while the other is the reaction in the reverse direction. Initially, the forward reaction occurs at a faster rate than the reverse reaction to form products. As the initial reactants are consumed, products accumulate. Then the forward reaction slows, and the rate of the reverse reaction increases.

Vocabulary

acid ['æsɪd]　　n. 酸
　　　　　　　　adj. 酸的
citric ['sɪtrɪk]　　adj. 柠檬的
equilibrium [ˌiːkwɪ'lɪbriəm]　　n. 平衡；均衡；均势
ascorbic [əs'kɔːbɪk]　　n. 抗坏血酸
hydrochloric /ˈhaɪdrəˈklɒrɪk/英式　　adj. 氯化氢的，盐酸的
antacid [ænt'æsɪd]　　n. 抗酸剂，解酸药
sodium ['səʊdiəm]　　n. 钠
bicarbonate [ˌbaɪ'kɑːbənət]　　n. 镁
corrosion [kə'rəʊʒn]　　n. 腐蚀
fluctuation [ˌflʌktʃu'eɪʃn]　　n. 波动
indicator ['ɪndɪkeɪtə]　　n. 指示物；标志；指示器；显示器
corrode [kə'rəʊd]　　v. 腐蚀；侵蚀；被腐蚀；被侵蚀；逐渐损害
nonmetal [ˌnɒn'metəl]　　n. 非金属
hydrogen ['haɪdrədʒən]　　n. 氢
polyatomic [ˌpɒlɪə'tɒmɪk]　　adj. 多原子的
electrolyte [ɪ'lektrəlaɪt]　　n. 电解质；电解液
chloride ['klɔːraɪd]　　n. 氯化物
anion ['ænaɪən]　　n. 阴离子
litmus ['lɪtməs]　　n. 石蕊
definition [ˌdefɪ'nɪʃn]　　n. 解释；定义；清晰度
slippery ['slɪpəri]　　adj. 滑得抓不住（或站不稳、难以行走）的
transfer [træns'fɜː(r)]　　v.（使）转移，搬迁；（使）调动；转职
ammonia [ə'məʊniə]　　n. 氨；氨水
nitrogen ['naɪtrədʒən]　　n. 氮；氮气
hydronium [haɪ'drəʊniəm]　　n. 水合氢（离子）
donate [dəʊ'neɪt]　　v. 捐赠；赠送
ionization [ˌaɪənaɪ'zeɪʃn]　　n. 电离，电子化

equation [ɪ'kweɪʒn]　　*n*. 方程；方程式；等式；相等；等同
molecule ['mɒlɪkjuːl]　　*n*. 分子
dissociate [dɪ'səʊʃieɪt]　　*v*. 解离；否认同……有关系；把……分开
aqueous ['eɪkwiəs]　　*adj*. 水的；含水的；水状的
equilibrium [ˌiːkwɪ'lɪbriəm]　　*n*. 平衡；均衡；均势
reversible [rɪ'vɜːsəbl]　　*adj*. 可逆的；可恢复原状的
consume [kən'sjuːm]　　*v*. 消耗；耗费（燃料、能量、时间等）
accumulate [ə'kjuːmjəleɪt]　　*v*. 积累；积聚

Part 1 Elements and Inorganic Compounds

Lesson 7 Reactions of Acids and Bases

Typical reactions of acids and bases include the reactions of acids with metals, bases, and carbonate or bicarbonate ions. For example, when you drop an antacid tablet in water, the bicarbonate ion and citric acid in the tablet react to produce carbon dioxide bubbles, water, and a salt. A salt is an ionic compound that does not have H^+ as the cation or OH^- as the anion.

Acids and Metals

Acids react with certain metals to produce hydrogen gas (H_2) and a salt. Metals that react with acids include potassium, sodium, calcium, magnesium, aluminum, zinc, iron, and tin.

In reactions that are single replacement reactions, the metal ion replaces the hydrogen in the acid.

$$Mg(s) + 2HCl(aq) \longrightarrow H_2(g) + MgCl_2(aq)$$

$$Zn(s) + 2HNO_3(aq) \longrightarrow H_2(g) + Zn(NO_3)_2(aq)$$

Acids and Hydroxides: Neutralization

Neutralization is a reaction between an acid and a base to produce water and a salt. The H^+ of an acid that can be strong or weak and the OH^- of a strong base combine to form water as one product. The salt is the cation from the base and the anion from the acid. We can write the following equation for the neutralization reaction between HCl and NaOH:

$$HCl(aq) + NaOH(aq) \longrightarrow H_2O(l) + NaCl(aq)$$

In a neutralization reaction, one H^+ always combines with one OH^-. Therefore, a neutralization equation uses coefficients to balance H^+ in the acid with the OH^- in the base.

Acid-Base Titration

Suppose we need to find the molarity of a solution of HCl, which has an unknown concentration. We can do this by a laboratory procedure called titration in which we neutralize an acid sample with a known amount of base. In a titration, we place a measured volume of the acid in a flask and add a few drops of an indicator such as phenolphthalein. In an acidic solution, phenolphthalein is colorless. Then we fill a buret with a NaOH solution of known molarity and carefully add NaOH solution to the acid in the flask. In the titration, we neutralize the acid by adding a volume of base that contains a matching number of moles of OH^-. We know that neutralization has taken place when the phenolphthalein in the solution changes from colorless to pink. This is called the neutralization endpoint. From the volume added and molarity of the NaOH solution, we can calculate the number of moles of NaOH, the moles of acid, and then the concentration of the acid.

Acid and Basic Salts

It is conceivable that in the neutralization of an acid by a base, only a part of the hydrogen might be neutralized; thus

$$NaOH + H_2SO_4 \longrightarrow NaHSO_4 + H_2O$$

The compound $NaHSO_4$ has acid properties, since it contains hydrogen, and is also a salt, since it contains both a metal and an acid radical. Such a salt containing acidic hydrogen is termed an acid salt. Phosphoric acid (H_3PO_4) might be progressively neutralized to form the salts, NaH_2PO_4, Na_2HPO_4, and Na_3PO_4. The first two are acid salts, since they contain replaceable hydrogen. A way of naming these salts is to call Na_2HPO_4 disodium hydrogen phosphate and NaH_2PO_4, sodium di-hydrogen phosphate. These acid phosphates are important in controlling the alkalinity of the blood. The third compound, sodium phosphate Na_3PO_4, which contains no replaceable hydrogen, is often referred to as normal sodium phosphate, or trisodium phosphate. to differentiate it from the two acid salts.

Historically, the prefix bi- has been used in naming some acid salts; in industry, for example, $NaHCO_3$ is called sodium bicarbonate and $Ca(HSO_3)_2$ calcium bisulfite. Since the bi- is somewhat misleading, the system of naming discussed above is preferable.

If the hydroxyl radicals of a base are progressively neutralized by an acid, basic salts may be formed:

$$Ca(OH)_2 + HCl \longrightarrow Ca(OH)Cl + H_2O$$

Basic salts have properties of a base and will react with acids to form a normal salt and water. The OH group in a basic salt is called an hydroxy group. The name of $Bi(OH)_2NO_3$ would be bismuth dihydroxynitrate. If the hydrogen atoms in an acid are replaced by two or more different metals, a mixed salt results. Thus the two hydrogen atoms in H_2SO_4 may be replaced with sodium and potassium to yield the mixed salt $NaKSO_4$, sodium potassium sulfate. $Na(NH_4)HPO_4$ is a mixed acid salt that may be crystallized from urine.

Part 1 Elements and Inorganic Compounds

Vocabulary

typical ['tɪpɪkl]   *adj.* 典型的；有代表性的
carbonate ['kɑːbənət]   *n.* 碳酸盐
bicarbonat [ˌbaɪ'kɑːbənət]   *n.* 碳酸氢盐
antacid [ænt'æsɪd]   *n.* 抗酸剂
citric ['sɪtrɪk]   *n.* 柠檬酸
bubbles ['bʌblz]   *n.* 气泡；泡沫
neutralization [ˌnjuːtrəlaɪ'zeɪʃ(ə)n]   *n.* 中和；中和作用
coefficients [ˌkəʊɪ'fɪʃənts]   *n.* 系数；率
molarity [məʊ'lærɪtɪ]   *n.* 摩尔浓度
concentration [ˌkɒnsn'treɪʃn]   *n.* 浓度；含量
procedure [prə'siːdʒə]   *n.* 程序；步骤；手续
titration [tɪ'treɪʃn]   *n.* 滴定；滴定法
neutralize ['njuːtrəlaɪz]   *v.* 使（酸性或碱性物质）中和
measured ['meʒəd]   *adj.* 有节奏的；一定量的
flask [flɑːsk]   *n.* 容器；瓶子；热水瓶；细颈烧瓶
phenolphthalein [ˌfiːnɒl'(f)θæliːn]   *n.* 酚酞
buret [bjʊ'ret]   *n.* 滴定管；玻璃量管
molarity [məʊ'lærɪtɪ]   *n.* 摩尔浓度
flask [flɑːsk]   *n.* 烧瓶；长颈瓶
neutralize ['njuːtrəlaɪz]   *v.* 中和；使成为中性
volume ['vɒljuːm]   *n.* 体积，容积，总数，总量
concentration [ˌkɒns(ə)n'treɪʃ(ə)n]   *n.* 含量，浓度
conceivable [kən'siːvəb(ə)l]   *adj.* 可以想象的
salt [sɔːlt]   *n.* 盐
radical ['rædɪk(ə)l]   根；游离基；自由基
phosphoric acid [fɒsˌfɒrɪk 'æsɪd]   磷酸
neutralized ['njuːtrəlaɪzd]   *v.* 使无效；中和；使成为中性
sodium dihydrogen phosphate ['səʊdɪəm] [dɪhaɪdrədʒən] [fɒsfeɪt]   磷酸二氢钠
disodium hydrogen phosphate ['dɪsəʊdɪəm] [haɪdrədʒən] [fɒsfeɪt]   磷酸氢二钠
sodium phosphate ['səʊdɪəm] ['fɒsfeɪt]   磷酸钠

Part ❷
Various Kinds of Organic Compounds

Part 2 Various Kinds of Organic Compounds

Lesson 8 Alkanes and Cycloalkanes

More than 90% of the compounds in the world are organic compounds. The large number of carbon compounds is possible because the covalent bond between carbon atoms (C—C) is very strong, allowing carbon atoms to form long, stable chains.

Naming Alkanes

The alkanes are a type of hydrocarbon in which the carbon atoms are connected only by single bonds. One of the most common uses of alkanes is as fuels. Methane, used in gas heaters and gas cooktops, is an alkane with one carbon atom. The alkanes ethane, propane, and butane contain two, three, and four carbon atoms, respectively, connected in a row or a *continuous chain*.

As we can see, the names for alkanes end in *ane*. Such names are part of the IUPAC (International Union of Pure and Applied Chemistry) system used by chemists to name organic compounds. Alkanes with five or more carbon atoms in a chain are named using Greek prefixes: *pent* (5), *hex* (6), *hept* (7), *oct* (8), *non* (9), and *dec* (10) (see Table 8.1).

Table 8.1 IUPAC Names of the First 10 Alkanes

Number of Molecular	Carbon Atoms Prefix	Name	Formula	Condensed Structural Formula
1	Meth	Methane	CH_4	CH_4
2	Eth	Ethane	C_2H_6	$CH_3—CH_3$
3	Prop	Propane	C_3H_8	$CH_3—CH_2—CH_3$
4	But	Butane	C_4H_{10}	$CH_3—CH_2—CH_2—CH_3$
5	Pent	Pentane	C_5H_{12}	$CH_3—CH_2—CH_2—CH_2—CH_3$
6	Hex	Hexane	C_6H_{14}	$CH_3—CH_2—CH_2—CH_2—CH_2—CH_3$
7	Hept	Heptane	C_7H_{16}	$CH_3—CH_2—CH_2—CH_2—CH_2—CH_2—CH_3$
8	Oct	Octane	C_8H_{18}	$CH_3—CH_2—CH_2—CH_2—CH_2—CH_2—CH_2—CH_3$
9	Non	Nonane	C_9H_{20}	$CH_3—CH_2—CH_2—CH_2—CH_2—CH_2—CH_2—CH_2—CH_3$
10	Dec	Decane	$C_{10}H_{22}$	$CH_3—CH_2—CH_2—CH_2—CH_2—CH_2—CH_2—CH_2—CH_2—CH_3$

When an alkane has four or more carbon atoms, the atoms can be arranged so that a side group called a *branch* or substituent is attached to a carbon chain. For example, there are different ball-and-stick models for two compounds that have the molecular formula C_4H_{10}. One model is shown as a chain of four carbon atoms. In the other model, a carbon atom is attached as a branch or substituent to a carbon in a chain of three atoms(see below). An alkane with at least one branch is called a *branched alkane*. When the two compounds have the same molecular formula but different arrangements of atoms, they are called structural isomers.

$$CH_3-CH_2-CH_2-CH_3 \qquad CH_3-\underset{\underset{CH_3}{|}}{\overset{\overset{CH_3}{|}}{CH}}-CH_3$$

In the IUPAC names for alkanes, a carbon branch is named as an alkyl group, which is an alkane that is missing one hydrogen atom. The alkyl group is named by replacing the *ane* ending of the corresponding alkane name with *yl*. Alkyl groups cannot exist on their own: They must be attached to a carbon chain. When a halogen atom is attached to a carbonchain, it is named as a *halo* group: *fluoro*, *chloro*, *bromo*, or *iodo*. Some of the common groups attached to carbon chains are illustrated in Table 8.2.

Table 8.2 Names and Formulas for Some Common Substituents

Substituent	Name
CH_3-	Methyl
CH_3-CH_2-	Ethyl
$CH_3-CH_2-CH_2-$	Propyl
$CH_3-\overset{\overset{\|}{}}{CH}-CH_3$	Isopropyl
$F-$	Fluoro
$Cl-$	chloro
$Br-$	bromo
$I-$	iodo

In the IUPAC system of naming, a carbon chain is numbered to give the location of the substituents.

Examples of naming alkanes:

$$CH_3-\overset{\overset{CH_3}{|}}{CH}-\underset{\underset{CH_3}{|}}{CH}-CH_3 \qquad\qquad CH_3-\overset{\overset{CH_3}{|}}{CH}-CH_2-\underset{\underset{CH_3}{|}}{\overset{\overset{Br}{|}}{C}}-CH_2-CH_3$$

2, 3-Dimethylbutane   4-bromo-2, 4-dimethylhexane

Part 2 Various Kinds of Organic Compounds

Naming Cycloalkanes

Hydrocarbons can also form cyclic or ring structures called cycloalkanes, which have two fewer hydrogen atoms than the corresponding alkanes. The simplest cycloalkane, cyclopropane (C_3H_6), has a ring of three carbon atoms bonded to six hydrogen atoms. Most often cycloalkanes are drawn using their skeletal formulas, which appear as simple geometric figures. A cycloalkane is named by adding the prefix *cyclo* to the name of the alkane with the same number of carbon atoms.

When one substituent is attached to a carbon atom in a cycloalkane, the name of the substituent is placed in front of the cycloalkane name. No number is needed for a single alkyl group or halogen atom because the carbon atoms in the cycloalkane are equivalent.

Examples of naming cycloalkanes:

Methylcyclopentane

Properties: Stability

Alkanes are relatively inert, chemically, since they are indifferent to reagents which react readily with alkenes or with alkynes. *n*-Hexane, for example, is not attacked by concentrated sulfuric acid, boiling nitric acid, molten sodium hydroxide, potassium permanganate, or chromic acid; with the exception of sodium hydroxide, these reagents all attack alkenes at room temperature. The few reactions of which alkanes are capable require a high temperature or special catalysis.

Properties: Halogenation

If a test tube containing *n*-hexane is put in a dark place and treated with a drop of bromine, the original color will remain undiminished in intensity for days. If the solution is exposed to sunlight, the color fades in a few minutes, and breathing across the mouth of the tube produces a cloud of condensate revealing hydrogen bromide as one reaction product. The reaction is a photochemical substitution:

$$C_6H_{14} + Br_2 \xrightarrow{Light} C_6H_{13}Br + HBr$$

Chlorination of alkanes is more general and more useful than bromination and can be effected not only photochemically but also by other methods.

Properties: Cracking

Heated to temperatures in the range 500-700 °C, higher alkanes undergo pyrolytic rupture or cracking to mixtures of smaller molecules, some saturated and some unsaturated. Unsaturated hydrocarbons produced by selective cracking of specific petroleum fractions are useful in chemical synthesis. Cracking ruptures carbon-carbon rather than carbon-hydrogen bonds because the energy required to break the C—C bond is 247 kJ/mol or kJ·mol^{-1}, whereas the C—H bond energy is 364 kJ/mol or kJ·mol^{-1}.

Properties: Oxidation

The reaction of hydrocarbons with oxygen with the output of energy is the basis for use of gasoline as fuel in internal combustion engines. The energy release on burning a given hydrocarbon is expressed as the heat of combustion in terms of kJ/mol or kJ·mol^{-1}.

Incomplete combustion of gaseous hydrocarbons is important in the manufacture of carbon blacks, particularly lampblack, a pigment for ink, and channel black, used as a filler in rubber compounding. Natural gas is used because of its cheapness and availability; the yield of black varies with the type of gas and the manufacturing process but usually is in the range of 2%-6% of the theoretical amount.

Partial air oxidation of a more limited extent is a means for production of specific oxygenated substances. Controlled air oxidation of high-boiling mineral oils and waxes from petroleum affords mixtures of higher carboxylic acids similar to those derived from fats and suitable for use in making soaps.

Vocabulary

alkane ['ælkein] *n.* 烷烃

hydrocarbon ['haidrou'kɑ:bən] *n.* 烃，碳氢化合物

pentane ['pentein] *n.* 戊烷

methane ['meθein] *n.* 甲烷

ethane ['eθein] *n.* 乙烷

propane ['prəpein] *n.* 丙烷

butane ['bju:tein] *n.* 丁烷

alkene ['ælki:n] *n.* 烯烃

alkyne ['ælkain]　　n. 炔烃

attack [ə'tæk]　　v. 起（化学）反应；侵蚀；进攻

concentrate ['kɔnsentreit]　　v. 提浓；浓缩

sodium hydroxide　氢氧化钠

catalysis [kə'tælisis]　　n. 催化（作用）

petroleum [pi'trəuljəm]　　n. 石油

hexane ['heksein]　　n. 己烷

heptane ['heptein]　　n. 庚烷

content ['kɔntent]　　n. 含量

octane ['ɔktein]　　n. 辛烷

nonane ['nɔnein]　　n. 壬烷

decane ['dekein]　　n. 癸烷

increment ['inkrimənt]　　n. 增量；增加

homologous [hɔ'mɔləgəs]　　adj. 同系列的

homolog ['hɔmələg]　　n. 同系物

halogenation [hælədʒə'neiʃən]　　n. 卤化（作用）

test tube [test tju:b]　　试管

intensity [in'tensiti]　　n. 强度

fade [feid]　　v. 褪色

condensate [kən'denseit]　　n. 冷凝物（液）

reveal [ri'vi:l]　　vt. 呈现；展现

hydrogen bromide ['haidrədʒən'brəumaid]　　n. 溴化氢

photochemical [,fəutəu'kemikəl]　　adj. 光化学的

chloroalkane [,klɔ(:) rə'ælkein]　　n. 氯代烷

regenerate [ri'dʒenəreit]　　vt. 再生

efficiency [i'fiʃənsi]　　n. 效率

collision [kə'liʒən]　　n. 碰撞

radiation [,reidi'eiʃən]　　n. 辐射；照射；放射线

exothermic [,eksəu'θə:mik]　　adj. 放热的

explosively [iks'plousivli]　　adv. 爆炸式地

pyrolytic [,paiərə'litik]　　adj. 热解的；高温分解的

rupture ['rʌptʃə]　　n. v. 裂开；断裂

crack [kræk]   v. 裂解，断裂

fraction ['frækʃən]   n. 馏分；部分；分数

synthesis ['sinθisis]   (复 syntheses ['sinθisi:z])   n. 合成；综合

gasoline ['gæsəli:n]   n. 汽油

fuel [fjuəl]   n. 燃料

internal combustion engine   内燃机

combustion [kəm'bʌstʃən]   n. 燃烧

gaseous ['geizjəs]   adj. 气体的；气态的

carbon black   炭黑

lampblack ['læmp'blæk]   n. 灯黑

pigment ['pigmənt]   n. 颜料，色料

channel ['tʃænl]   black   槽法炭黑

compounding [kəm'paundiŋ]   n. 配料；配方

oxygenate [ɔk'sidʒineit]   vt. 氧化

fat [fæt]   n. 脂肪

Part 2　Various Kinds of Organic Compounds

Lesson 9　Alkenes and Alkynes

Alkenes and alkynes are families of hydrocarbons that contain double and triple bonds, respectively. They are called unsaturated hydrocarbons because they do not contain the maximum number of hydrogen atoms, as do alkanes. They react with hydrogen gas to increase the number of hydrogen atoms to become alkanes, which are saturated hydrocarbons.

Identifying Alkenes and Alkynes

Alkenes contain one or more carbon-carbon double bonds that form when adjacent carbon atoms share two pairs of valence electrons. Recall that a carbon atom always forms four covalent bonds. In the simplest alkene, ethene, C_2H_4, two carbon atoms are connected by a double bond and each is also attached to two H atoms. This gives each carbon atom in the double bond a trigonal planar arrangement with bond angles of 120°. As a result, the ethene molecule is flat because the carbon and hydrogen atoms all lie in the same plane.

Ethene, more commonly called ethylene, is an important plant hormone involved in promoting the ripening of fruit. Commercially grown fruit, such as avocados, bananas, and tomatoes, are often picked before they are ripe. Before the fruit is brought to market, it is exposed to ethylene to accelerate the ripening process. Ethylene also accelerates the breakdown of cellulose in plants, which causes flowers to wilt and leaves to fall from trees.

In an alkyne, a triple bond forms when two carbon atoms share three pairs of valence electrons. In the simplest alkyne, ethyne (C_2H_2), the two carbon atoms of the triple bond are each attached to one hydrogen atom, which gives a triple bond a linear geometry. Ethyne, commonly called acetylene, is used in welding where it reacts with oxygen to produce flames with temperatures above 3300 °C.

Naming Alkenes and Alkynes

The IUPAC names for alkenes and alkynes are similar to those of alkanes. Using the alkane name with the same number of carbon atoms, the ane ending is replaced with ene for an alkene and yne for an alkyne.

Examples of naming an alkene and an alkyne:

$$CH_3-CH(CH_3)-CH_2-CH=CH_2 \qquad CH_3-CH_2-CH_2-C\equiv C-CH_3$$
$$\text{4-methyl-2-pentene} \qquad\qquad\qquad \text{2-hexyne}$$

Physical Properties

Alkenes are known also as ethylenic hydrocarbons and as olefins. The term olefin, meaning oil-forming, was applied by early chemists because the gaseous members of the series combine with chlorine and bromine to form oily addition products.

Alkenes are hardly distinguishable from the corresponding saturated hydrocarbons. The boiling points are no more than a few degrees below those of alkanes of slightly higher molecular weight, and the densities are a few percent higher; in the first few members of the two series there is even a marked correspondence in the melting points. Cycloalkanes differ more from alkanes than alkenes do, and hence ring formation influences physical properties more than introduction of an ethylene linkage. The heat of combustion of 1-hexene is practically the same as that of *n*-hexane on either a weight or volume basis.

Addition Reactions

The most characteristic reaction of alkenes is the addition of atoms or groups of atoms to the carbon atoms in a double bond. Addition occurs because double bonds are easily broken, providing electrons to form new single bonds. The addition reactions have different names that depend on the type of reactant we add to the alkene, as Table 9.1 shows.

Table 9.1 Summary of Addition Reactions

Name of Addition Reaction	Reactants	Catalysts	Product
Hydrogenation	Alkene + H_2	Pt, Ni, or Pd	Alkane
Hydration	Alkene + H_2O	H^+ (strong acid)	Alcohol

Hydrogenation

In a reaction called hydrogenation, H atoms add to each of the carbon atoms in a double bond of an alkene. During hydrogenation, the double bonds are converted to single bonds in alkanes. A catalyst such as finely divided platinum (Pt), nickel (Ni), or palladium (Pd) is used to speed up the reaction.

Part 2 Various Kinds of Organic Compounds

Hydration

In hydration, an alkene reacts with water (H—OH). A hydrogen atom (H) from water forms a bond with one carbon atom in the double bond, and the oxygen atom in —OH forms a bond with the other carbon. The reaction is catalyzed by a strong acid such as H_2SO_4. Hydration is used to prepare alcohols, which have the hydroxyl (—OH) functional group. In the general equation for hydration, water is written as H—OH and the acid catalyst is represented by H^+.

When water adds to a double bond in which the carbon atoms are attached to different numbers of H atoms (an asymmetrical double bond), the H— from H—OH attaches to the carbon that has the *greater* number of H atoms and the —OH from H—OH adds to the other carbon atom from the double bond. In the following example, the H— from H—OH attaches to the end carbon of the double bond, which has more hydrogen atoms, and the —OH adds to the middle carbon atom.

Vocabulary

ethylene ['eθili:n]　*n.* 乙烯；1, 2-亚乙基，乙撑（—CH_2CH_2—）

triple ['tripl]　*adj.* 三重的

triple bond　三键

acetylene [ə'setili:n]　*n.* 乙炔

ethylenic [,eθi'lenik]　*adj.* 烯的；乙烯的

olefin ['əuləfin]　*n.* 烯烃

oil-forming [ɔil'fɔ:miŋ]　*n.* 生成石油；成油

oily ['ɔili]　*adj.* 油的；油状的；含油的

addition [ə'diʃən]　*n.* 加成；加合

distinguishable [dis'tiŋgwiʃəbl]　*adj.* 可区别的；可辨别的

density ['densiti]　*n.* 密度；比重

cycloalkane [,saikləu'ælkein]　*n.* 环烷（烃）

influence ['influəns]　*n.* 影响

linkage ['liŋkidʒ]　*n.* 键（合）

heat of combustion 燃烧热

volum ['vɔljum]　*n.* 卷；册；体积

synthetic [sin'θetik]　*adj.* 合成的；综合的

hydration [hai'dreiʃən]　*n.* 水化

eliminate [i'limineit]　*vt.* 消去；消除

ethanol ['eθənɔl]　*n.* 乙醇

distil [dis'til]　*v.* 蒸馏

Part 2 Various Kinds of Organic Compounds

Lesson 10 Alcohols, Aldehydes and Ketones

Alcohols, Phenols, and Ethers

In an alcohol, the functional group known as a hydroxyl group (—OH) replace hydrogen atom in a hydrocarbon. In a phenol, the hydroxyl group replaces a hydrogen atom attached to a benzene ring. In an ether, the functional group consists of an oxygen atom, which is attached to two carbon atoms (—O—). Molecules of alcohols, phenols, and ethers have bent shapes around the oxygen or sulfur atom, similar to water.

Naming Alcohols

In the IUPAC system, an alcohol is named by replacing the *e* of the corresponding alkane name with *ol*. The common name of a simple alcohol uses the name of the alkyl group followed by *alcohol*.

When an alcohol consists of a chain with three or more carbon atoms, the chain is numbered to give the position for the —OH group and any substituents on the chain. A cyclic alcohol is named as a *cycloalkanol*. If there are substituents, the ring is numbered from carbon 1, which is the carbon attached to the —OH group. Compounds with no substituents on the ring do not require a number for the hydroxyl group.

Naming Phenols

The term *phenol* is the IUPAC name for a benzene ring bonded to a hydroxyl group (—OH), which is used in the name of the family of organic compounds derived from phenol. When there is a second substituent, the benzene ring is numbered starting from carbon 1, which is the carbon bonded to the —OH group.

Phenol

3-Chlorophenol

Naming Ethers

An ether contains an oxygen atom that is attached by single bonds to two carbon groups that are alkyls or aromatic rings. Ethers have a bent structure like water and alcohols, dimethyl ether except both hydrogen atoms are replaced by carbon groups.

Most ethers have common names. The name of each alkyl or aromatic group attached to the oxygen atom is written in alphabetical order, followed by the word *ether*. In this text, we will use only the common names of ethers.

Aldehydes and Ketones

Aldehydes and ketones contain a carbonyl group that consists of a carbon-oxygen double bond with two groups of atoms attached to the carbon at angles of 120°. The double bond in the carbonyl group is similar to that of alkenes, except the carbonyl group has a dipole. The oxygen atom with two lone pairs of electrons is much more electronegative than the carbon atom. Therefore, the carbonyl group has a strong dipole with a partial negative charge 1d − 2 on the oxygen and a partial positive charge 1d + 2 on the carbon. The polarity of the carbonyl group strongly influences the physical and chemical properties of aldehydes and ketones.

In an aldehyde, the carbon of the carbonyl group is bonded to at least one hydrogen atom. That carbon may also be bonded to another hydrogen atom, a carbon of an alkyl group, or an aromatic ring. The aldehyde group may be written as separate atoms or as —CHO, with the double bond understood. In a ketone, the carbonyl group is bonded to two alkyl groups or aromatic rings. The keto group (C═O) can sometimes be written as CO. A skeletal formula may also be used to represent an aldehyde or ketone.

Naming Aldehydes

In the IUPAC system, an aldehyde is named by replacing the *e* of the corresponding alkane name with *al*. No number is needed for the aldehyde group because it always appears at the end of the chain. The aldehydes with carbon chains of one to four carbons are often referred to by their common names, which end in *aldehyde*. The roots (*form, acet, propion,* and *butyr*) of these common names are derived from Latin or Greek words.

IUPAC Methanal
Common (formaldehyde)

Ethanal
(acetaldehyde)

Propanal
(propionaldehyde)

Butanal
(butyraldehyde)

Naming Ketones

Aldehydes and ketones are some of the most important classes of organic compounds. Because they have played a major role in organic chemistry for more than a century, the common names for unbranched ketones are still in use. In the common names, the alkyl groups bonded to the carbonyl group are named as substituents and are listed alphabetically, followed by *ketone*. Acetone, which is another name for propanone, has been retained by the IUPAC system. In the IUPAC system, the name of a ketone is obtained by replacing the *e* in the corresponding alkane name with *one*. Carbon chains with five carbon atoms or more are numbered from the end nearer the carbonyl group.

$H_3C-C(=O)-CH_3$
Propanone
(dimethyl ketone; acetone)

$H_3C-CH_2-C(=O)-CH_3$
Butanone
(ethyl methyl ketone)

$H_3C-CH_2-C(=O)-CH_2-CH_3$
3-Pentanone
(diethyl ketone)

In a cyclic ketone, the carbonyl carbon is numbered as carbon 1. The ring is numbered in the direction to give substituents the lowest possible numbers.

3-Methylcyclohexanone

Reactions of Alcohols, Aldehydes, and Ketones

Alcohols, similar to hydrocarbons, undergo combustion in the presence of oxygen. For example, in a restaurant, a flaming dessert may be prepared by pouring liquor on fruit or ice cream and lighting it.

Dehydration of Alcohols to Form Alkenes

In a dehydration reaction, alcohols lose a water molecule when they are heated with an acid catalyst such as H_2SO_4. The components H— and —OH are removed from *adjacent carbon atoms of the same alcohol* to produce a water molecule. A double bond forms between the same two carbon atoms to produce an alkene product.

$$\text{H}-\underset{\underset{\text{H}}{|}}{\overset{\overset{\text{H}}{|}}{\text{C}}}-\underset{\underset{\text{H}}{|}}{\overset{\overset{\text{OH}}{|}}{\text{C}}}-\text{H} \xrightarrow[\text{heat}]{H^+} \underset{\text{H}}{\overset{\text{H}}{>}}\text{C}=\text{C}\underset{\text{H}}{\overset{\text{H}}{<}} + H_2O$$

Ethanol → Ethene

Cyclopentanol $\xrightarrow[\text{heat}]{H^+}$ Cyclopentene + H_2O

Oxidation of Alcohols

In organic chemistry, an oxidation involves the addition of oxygen or a loss of hydrogen atoms. As a result, there is an increase in the number of carbon-oxygen bonds. In a reduction reaction, the product has fewer bonds between carbon and oxygen. The oxidation of a primary alcohol produces an aldehyde, which contains a double bond between carbon and oxygen. For example, the oxidation of methanol and ethanol occurs by removing two hydrogen atoms, one from the —OH group and another from the carbon that is bonded to the —OH group. To indicate the presence of an oxidizing agent, the symbol [O] is written over the arrow to indicate that O in the H_2O product is obtained from the oxidizing agent. The oxidized product contains the same number of carbon atoms as the reactant.

In the oxidation of secondary alcohols, the products are ketones. Two hydrogen atoms are removed, one from the —OH group and another from the carbon bonded to the —OH group. The result is a ketone that has the carbon-oxygen double bond attached to alkyl groups on both sides. There is no further oxidation of a ketone because there are no hydrogen atoms attached to the carbonyl group.

Part 2　Various Kinds of Organic Compounds

Tertiary alcohols do not oxidize readily because there is no hydrogen atom on the carbon bonded to the —OH group. Because C—C bonds are usually too strong to oxidize, tertiary alcohols resist oxidation.

Oxidation of Aldehydes

Earlier in this section, we saw that primary alcohols can oxidize to aldehydes. Aldehydes oxidize further by the addition of another O to form a carboxylic acid, which has a *carboxyl* functional group. This step occurs so readily that it is often difficult to isolate the aldehyde product during the oxidation reaction. In contrast, ketones produced by the oxidation of secondary alcohols do not undergo further oxidation.

$$CH_3-CH_2-OH \xrightarrow{Oxidation} CH_3-\overset{\overset{O}{\|}}{C}-H \xrightarrow{Further\ oxidation} CH_3-\overset{\overset{O}{\|}}{C}-OH$$

　　Ethanol(1°)　　　　　　　　　Ethanal　　　　　　　　　　Ethanoic acid

$$CH_3-\overset{\overset{OH}{|}}{CH}-CH_3 \xrightarrow{Oxidation} CH_3-\overset{\overset{O}{\|}}{C}-CH_3 \xrightarrow{Further\ oxidation} no\ reaction$$

　　2-Propanol(2°)　　　　　　　Propanone

Reduction of Aldehydes and Ketones

Aldehydes and ketones are reduced by sodium borohydride ($NaBH_4$) or hydrogen (H_2). In the reduction of organic compounds, there is a decrease in the number of carbon-oxygen bonds by the addition of hydrogen or the loss of oxygen. Aldehydes reduce to primary alcohols, and ketones reduce to secondary alcohols. A catalyst such as nickel, platinum, or palladium is needed for the addition of hydrogen to the carbonyl group.

$$H_3C-CH_2-\overset{\overset{O}{\|}}{C}-H + H_2 \xrightarrow{Pt} H_3C-CH_2-\overset{\overset{OH}{|}}{\underset{\underset{H}{|}}{C}}-H$$

　　　　Propanal　　　　　　　　　　1-Propanol

Vocabulary

alcohol ['ælkəhɔl]　*n.* 酒精

functional ['fʌŋkʃənl]　*n.* 官能团

hydrocarbon [ˌhaidrəu'ka:bən]　*n.* 碳氢化合物，烃

phenol ['fi:nɔl]   n. 酚，苯酚
benzene [ben'zi:n]   n. 苯
sulfur ['sʌlfə]   n. 硫
substitutens ['sʌbstɪtju:t]   n. 取代
hydroxyl [hai'drɒksi]   n. 羟（基）；羟基；氢氧基
carbonyl ['kɑ:bənil]   n. 羰基
dipole ['daɪpəʊl]   n. 偶极-偶极相互作用
polarity [pə'lærəti]   n. 极性
aldehyde ['ældɪhaɪd]   n. 醛；乙醛
carbonyl ['kɑ:bənɪl]   n. 碳酰基；羰基
bonded ['bɒndɪd]   v. 键
alkyl ['ælkil]   n. 烷基；烃基
　　　　　　　adj. 烷基的；烃基
combustion [kəm'bʌstʃən]   n. 燃烧；燃烧过程
liquor ['lɪkər]   n. 酒；烈性酒；含酒精饮料
dehydration [ˌdi:haɪ'dreɪʃən]   v. 脱水
alkenes ['æl,kinz]   n. 烯烃；烯烃类；烯类
catalyst ['kætəlɪst]   n. 催化剂；促使变化的人；引发变化的因素
adjacent [ə'dʒeɪsnt]   adj. 相邻；邻近的；与……毗连的
oxidation [ˌɒksɪ'deɪʃn]   n. 氧化
tertiary alcohols ['tɜ:ʃəri] ['ælkəhɒl]   叔醇
oxidize ['ɒksɪdaɪz]   vt. 使氧化；使生锈
carboxylic [kɑ:'bɒksɪlɪk]   n. [化] 羧酸
carboxyl [kɑ:'bɒksɪl]   n. 羧基
ketone ['ki:təʊn]   n. 酮
undergo [ʌndə'gəʊ]   vt. 经历；经受；遭受
borohydride [bʌrəʊ'haɪdraɪd]   n. 硼氢化物
palladium [pə'leɪdiəm]   n. 钯

Lesson 11 Carboxylic Acids, Esters, Amines and Amides

Carboxylic acids are weak acids. They have a sour or tart taste, produce hydronium ions in water, and neutralize bases. You encounter carboxylic acids when you use a salad dressing containing vinegar, which is a solution of acetic acid and water, or experience the sour taste of citric acid in a grapefruit or lemon. When a carboxylic acid reacts with an alcohol, an ester and water are produced. Fats and oils are esters of glycerol and fatty acids, which are long-chain carboxylic acids. Esters produce the pleasant aromas and flavors of many fruits, such as bananas, strawberries, and oranges.

Amines and amides are organic compounds that contain nitrogen. Many nitrogen-containing compounds are important to life as components of amino acids, proteins, and nucleic acids (DNA and RNA). Many amines that exhibit strong physiological activity are used in medicine as decongestants, anesthetics, and sedatives. Examples include dopamine, histamine, epinephrine, and amphetamine.

Alkaloids such as caffeine, nicotine, cocaine, and digitalis, which have powerful physiological activity, are naturally occurring amines obtained from plants. In an amide, the functional group consists of a carbonyl group attached to an amine. Amides, which are derived from carboxylic acids, are important in biology in proteins. In biochemistry, the amide bond that links amino acids in a protein is called a peptide bond. Some medically important amides include acetaminophen (Tylenol) used to reduce fever; phenobarbital, a sedative and anticonvulsant medication; and penicillin, an antibiotic.

Carboxylic Acids

In a carboxylic acid, the carbon atom of a carbonyl group is attached to a hydroxyl group that forms a carboxyl group. The carboxyl functional group may be attached to an alkyl group or an aromatic group.

IUPAC Names of Carboxylic Acids

The IUPAC names of carboxylic acids replace the *e* of the corresponding alkane name with *oic acid*. If there is a substituent, the carbon chain is numbered beginning with the carboxyl carbon.

Common Names of Carboxylic Acids

Many carboxylic acids are still named by their common names, which use prefixes: *form*, *acet*, *propion*, *butyr*. These prefixes are related to the natural sources of the simple carboxylic acids. For example, formic acid is injected under the skin from bee or red ant stings and other insect bites. Acetic acid is the oxidation product of the ethanol in wines and apple cider. A solution of acetic acid and water is known as vinegar. Butyric acid gives the foul odor to rancid butter. The following examples of some typical Carboxylic Acids.

Condensed Structural Formula	IUPAC Name	Common Name
H—C(=O)—OH	Methanoic acid	Formic acid
H_3C—C(=O)—OH	Ethanoic acid	Acetic acid
H_3C—CH_2—C(=O)—OH	Propanoic acid	Propionic acid
H_3C—CH_2—CH_2—C(=O)—OH	Butanoic acid	Butyric acid

Esters

A carboxylic acid reacts with an alcohol to form an ester and water. In an ester, the —H of the carboxylic acid is replaced by an alkyl group. Fats and oils in our diets contain esters of glycerol and fatty acids, which are long-chain carboxylic acids. The aromas and flavors of many fruits, including bananas, oranges, and strawberries, are due to esters.

Naming Esters

The name of an ester consists of two words, which are derived from the names of the alcohol and the acid in that ester. The first word indicates the *alkyl* part from the alcohol. The second word is the name of the *carboxylate* from the carboxylic acid. The IUPAC names

Part 2　Various Kinds of Organic Compounds

of esters use the IUPAC names of the acids, while the common names of esters use the common names of the acids. Let's take a look at the following ester, which has a pleasant, fruity odor. We start by separating the ester bond to identify the alkyl part from the alcohol and the carboxylate part from the acid. Then we name the ester as an alkyl carboxylate. The following examples of some typical esters show the IUPAC names, as well as the common names, of esters.

$$H_3C-\overset{\overset{O}{\|}}{C}-O-CH_2-CH_3$$
Ethyl ethanoate
(ethyl acetate)

$$H_3C-CH_2-\overset{\overset{O}{\|}}{C}-O-CH_3$$
Methyl propanoate
(methyl propionate)

$$C_6H_5-\overset{\overset{O}{\|}}{C}-O-CH_2-CH_3$$
Ethyl benzoate

Amines

Amines are derivatives of ammonia (NH_3), in which one or more hydrogen atoms are replaced with alkyl or aromatic groups. In methylamine, a methyl group replaces one hydrogen atom in ammonia. The bonding of two methyl groups gives dimethylamine, and the three methyl groups in trimethylamine replace all the hydrogen atoms in ammonia.

Naming and Classifying Amines

The common names of amines are often used when the alkyl groups bonded to the nitrogen atom are not branched. Then the alkyl groups are listed in alphabetical order. The prefixes *di* and *tri* are used to indicate two and three identical substituents. Amines are classified by counting the number of carbon atoms directly bonded to the nitrogen atom. In a primary (1°) amine, the nitrogen atom is bonded to one alkyl group. In a secondary (2°) amine, the nitrogen atom is bonded to two alkyl groups. In a tertiary (3°) amine, the nitrogen atom is bonded to three alkyl groups.

$$\overset{..}{H-N-H} \atop H$$
Ammonia

$$\overset{..}{H_3C-N-H} \atop H$$
Primary (1°) amine

$$\overset{..}{H_3C-N-CH_3} \atop H$$
Secondary (2°) amine

$$\overset{..}{H_3C-N-CH_3} \atop CH_3$$
Tertiary (3°) amine

Aromatic Amines

The aromatic amines use the name *aniline*, which is approved by IUPAC. Alkyl groups attached to the nitrogen of aniline are named with the prefix *N-* followed by the alkyl name.

Aniline   4-Bromoaniline   *N*-Methylaniline

Amides

The amides are derivatives of carboxylic acids in which a nitrogen group replaces the hydroxyl group.

Naming Amides

In both the IUPAC and common names, amides are named by dropping the *oic acid* (IUPAC) or *ic acid* (common) from the carboxylic acid name and adding the suffix *amide*. When alkyl groups are attached to the nitrogen atom, the prefix *N-* or *N, N-* precedes the name of the amide, depending on whether there are one or two groups. We can diagram the name of an amide in the following way:

Methanamide (formamide)   Ethanamide (acetamide)   *N*-Methylpropanamide (*N*-methylpropionamide)   Benzamide

Vocabulary

carboxylic [ˌkɑːbɔk'silik]   acids   羧酸；羧酸类物质
salad dressing   沙拉酱
fats and oils   脂肪与油类
citric ['sitrik]   acid   柠檬酸；柠檬醇
fatty [fæti]   acids   脂肪酸；脂肪酸成分
acetic [ə'siːtik]   acid   乙酸；醋酸
organic compounds   有机化合物

anticonvulsant [ˌæntikən'vʌlsənt] *v.* 抗惊厥；抗惊厥的
component [kəm'pəʊnənt] *n.* 成分
acetaminophen [əˌsiːtə'minəfen] *n.* 对乙酰氨基酚
nucleic acids 核酸；核酸分子
derived from 来源
amino acids 氨基酸
amide bond 酰胺键；酰胺结构
functional group 官能团；功能基
natural sources 自然来源；天然源
carboxyl group 羧基；羧酸基团；羧酸根
hydroxyl group 羟基；羟自由基；氢氧根
aromatic group 芳香烃；芳香族；芳香基团
butyric acid 丁酸；正丁酸
acetic acid 乙酸；醋酸
formic acid 甲酸；甲酸
ethanol ['eθənɒl] *n.* 乙醇
ester ['estə(r)] bond 酯键
alphabetical [ˌælfə'betikl] *n.* 字母的；按字母顺序的
bonded to 键合
prefix ['priːfiks] *n.* 前缀；前置词；字头；词头
amine [ə'miːn] *n.* 胺；胺的；胺类；胺类化合物
aromatic amines 芳香胺；芳香胺类化合物
aniline ['ænili:n] *n.* 阿尼林；氨基苯
mide ['æmaid] *n.* 酰胺类；酰胺；酰胺类化合物；氨基化合物

Part 3

Basic Knowledge of Materials Chemistry

Part 3 Basic Knowledge of Materials Chemistry

Lesson 12 Characteristics and Applications of Metals

Ferrous alloy

Ferrous alloys are those materials of which iron is the prime constituent. They are produced in larger quantities than any other metal type. They are especially important as engineering construction materials. Their widespread use is accounted for by three factors: (1) iron-containing compounds exist in abundant quantities within the earth's crust; (2) metallic iron and steel alloys may be produced using relatively economical extraction, refining, alloying, and fabrication techniques; and (3) ferrous alloys are extremely versatile, in that they may be tailored to have a wide range of mechanical and physical properties. The principal disadvantage of many ferrous alloys is their susceptibility to corrosion. This section discusses compositions, microstructures, and properties of a number of different classes of steels and cast irons. The classification scheme for the various ferrous alloys is shown in Figure 12.1.

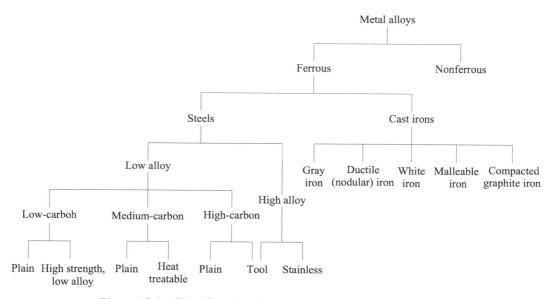

Figure 12.1 Classification scheme for the various ferrous alloys

Steel

Steels are iron-carbon alloys that may contain appreciable concentrations of other alloying elements. The mechanical properties are sensitive to the content of carbon, which is normally less than 1.0 wt%. The common steels are classified according to carbon concentration-namely, into low-, medium-, and highcarbon steel types.

Carbon Steels

Low-Carbon Steels

Of all the different steels, those produced in the greatest quantities fall within the low-carbon classification. These generally contain less than about 0.25 wt% C and are unresponsive to heat treatments intended to form martensite; strengthening is accomplished by cold work. Microstructures consist of ferrite and pearlite constituents. As a consequence, these alloys are relatively soft and weak but have outstanding ductility and toughness; in addition, they are machinable, weldable, and, of all steels, are the least expensive to produce. Typical applications include automobile body components, structural shapes, and sheets that are used in pipelines, buildings, bridges, and tin cans.

Medium-Carbon Steels

The medium-carbon steels have carbon concentrations between about 0.25 and 0.60 wt%. These alloys may be heat treated by austenitizing, quenching, and then tempering to improve their mechanical properties. They are most often utilized in the tempered condition, having microstructures of tempered martensite. The plain medium-carbon steels have low hardenabilities and can be successfully heat treated only in very thin sections and with very rapid quenching rates. Additions of chromium, nickel, and molybdenum improve the capacity of these alloys to be heat treated, giving rise to a variety of strength-ductility combinations. These heat-treated alloys are stronger than the low-carbon steels, but at a sacrifice of ductility and toughness. Applications include railway wheels and tracks, gears, crankshafts, and other machine parts and high-strength structural components calling for a combination of high strength, wear resistance, and toughness.

High-Carbon Steels

The high-carbon steels, normally having carbon contents between 0.60 and 1.4 wt%, are the hardest, strongest, and yet least ductile of the carbon steels. They are almost always used in a hardened and tempered condition and, as such, are especially wear resistant and capable of holding a sharp cutting edge. The tool and die steels are high-carbon alloys, usually containing chromium, vanadium, tungsten, and molybdenum. These alloying elements combine with carbon to form very hard and wear-resistant carbide compounds. These steels are utilized as cutting tools and dies for forming and shaping materials, as well as in knives, razors, hacksaw blades, springs, and high-strength wire.

Stainless Steels

The stainless steels are highly resistant to corrosion in a variety of environments, especially the ambient atmosphere. Their predominant alloying element is chromium; a concentration of at least 11 wt% Cr is required. Their corrosion resistance may also be enhanced by nickel and molybdenum additions. The stainless steels are divided into three classes on the basis of the predominant phase constituent of the microstructure—martensitic, ferritic, or austenitic. A wide range of mechanical properties combined with excellent resistance to corrosion make stainless steels very versatile in their applicability.

Cast Irons

Generically, cast irons are a class of ferrous alloys with carbon contents above 2.14 wt%; in practice, however, most cast irons contain between 3.0 and 4.5 wt% C and, in addition, other alloying elements. They are easily melted and amenable to casting. Furthermore, some cast irons are very brittle, and casting is the most convenient fabrication technique.

Gray Cast Iron

The carbon and silicon contents of gray cast irons vary between 2.5-4.0 wt% and 1.0-3.0 wt%, respectively. For most of these cast irons, the graphite exists in the form of flakes (similar to corn flakes), which are normally surrounded by an ferrite or pearlite matrix. Because of these graphite flakes, a fractured surface takes on a gray appearance, hence its name.

Nodular (Ductile) Iron

Adding a small amount of magnesium and/or cerium to the gray iron before casting produces a distinctly different microstructure and set of mechanical properties. Graphite still forms, but as nodules or sphere-like particles instead of flakes. The resulting alloy is called nodular or ductile iron.

White Cast Iron

For low-silicon cast irons (containing less than 1.0 wt% Si) and rapid cooling rates, most of the carbon exists as cementite instead of graphite. A fracture surface of this alloy has a white appearance, and thus it is termed white cast iron.

Compacted Graphite Iron

A relatively recent addition to the family of cast irons is compacted graphite iron (abbreviated CGI). As with gray, ductile, and malleable irons, carbon exists as graphite, which formation is promoted by the presence of silicon. Silicon content ranges between 1.7 and 3.0 wt%, whereas carbon concentration is normally between 3.1 and 4.0 wt%.

Nonferrous alloys

Steel and other ferrous alloys are consumed in exceedingly large quantities because they have such a wide range of mechanical properties, may be fabricated with relative ease, and are economical to produce. However, they have some distinct limitations, chiefly: (1) a relatively high density, (2) a comparatively low electrical conductivity, and (3) an inherent susceptibility to corrosion in some common environments. Thus, for many applications it is advantageous or even necessary to utilize other alloys having more suitable property combinations. Alloy systems are classified either according to the base metal or according to some specific characteristic that a group of alloys share. This section discusses the following metal and alloy systems: copper, aluminum, magnesium, and titanium alloys, the refractory metals, the superalloys and the noble metals.

Copper and Its Alloys

Copper and copper-based alloys, possessing a desirable combination of physical properties, have been utilized in quite a variety of applications since antiquity. Unalloyed copper is so soft and ductile that it is difficult to machine; also, it has an almost unlimited capacity to be cold worked. Furthermore, it is highly resistant to corrosion in diverse environments including the ambient atmosphere, seawater, and some industrial chemicals. The mechanical and corrosion-resistance properties of copper may be improved by alloying. Most copper alloys cannot be hardened or strengthened by heat-treating procedures; consequently, cold working and/or solid-solution alloying must be utilized to improve these mechanical properties. Copper and its Alloys mainly contain brass and bronze. The most common copper alloys are the brasses for which zinc, as a substitutional impurity, is the predominant alloying element. The bronzes are alloys of copper and several other elements, including tin, aluminum, silicon, and nickel. These alloys are somewhat stronger than the brasses, yet they still have a high degree of corrosion resistance.

Magnesium and Its Alloys

Perhaps the most outstanding characteristic of magnesium is its density 1.7 g/cm^3, which is the lowest of all the structural metals; therefore, its alloys are used where light weight is an important consideration (e.g., in aircraft components). Magnesium has a hexagonal close-packed crystal structure, is relatively soft, and has a low elastic modulus. At room temperature magnesium and its alloys are difficult to deform; in fact, only small degrees of cold work may be imposed without annealing. Magnesium, like aluminum, has a moderately low melting temperature (651 °C). Chemically, magnesium alloys are relatively unstable and especially susceptible to corrosion in marine environments. On the other hand, corrosion or oxidation resistance is reasonably good in the normal atmosphere; it is believed that this behavior is due to impurities rather than being an inherent characteristic of Mg alloys. Fine magnesium powder ignites easily when heated in air; consequently, care should be exercised when handling it in this state. Aluminum, zinc, manganese, and some of the rare earths are the major alloying elements. These alloys are also classified as either cast or wrought, and some of them are also heat treatable.

Titanium and Its Alloys

Titanium and its alloys are relatively new engineering materials that possess an extraordinary combination of properties. The pure metal has a relatively low density (4.5 g/cm^3), a high melting point, and an elastic modulus of 107 GPa. Titanium alloys are extremely strong; room temperature tensile strengths as high as 1400 MPa are attainable, yielding remarkable specific strengths. Furthermore, the alloys are highly ductile and easily forged and machined.

The major limitation of titanium is its chemical reactivity with other materials at elevated temperatures. This property has necessitated the development of nonconventional refining, melting, and casting techniques; consequently, titanium alloys are quite expensive. In spite of this high temperature reactivity, the corrosion resistance of titanium alloys at normal temperatures is unusually high; they are virtually immune to air, marine, and a variety of industrial environments. They are commonly utilized in airplane structures, space vehicles, surgical implants, and in the petroleum and chemical industries.

The Refractory Metals

Metals that have extremely high melting temperatures are classified as the refractory metals. Included in this group are niobium (Nb), molybdenum (Mo), tungsten (W), and tantalum (Ta). Melting temperatures range between 2468 °C for niobium and 3410 °C, the highest melting temperature of any metal, for tungsten. Interatomic bonding in these metals is extremely strong, which accounts for the melting temperatures, and, in addition, large elastic moduli and high strengths and hardnesses, at ambient as well as elevated temperatures. The applications of these metals are varied. For example, tantalum and molybdenum are alloyed with stainless steel to improve its corrosion resistance. Molybdenum alloys are utilized for extrusion dies and structural parts in space vehicles; incandescent light filaments, X-ray tubes, and welding electrodes employ tungsten alloys. Tantalum is immune to chemical attack by virtually all environments at temperatures below and is frequently used in applications requiring such a corrosion-resistant material.

The Superalloys

The alloys that have extremely excellent heat resistance are named as the superalloys. They have superlative combinations of properties. Most are used in aircraft turbine components, which must withstand exposure to severely oxidizing environments and high temperatures for reasonable time periods. Mechanical integrity under these conditions is critical; in this regard, density is an important consideration because centrifugal stresses are diminished in rotating members when the density is reduced. These materials are classified according to the predominant metal in the alloy, which may be cobalt, nickel, or iron. Other alloying elements include the refractory metals (Nb, Mo, W, Ta), chromium, and titanium. In addition to turbine applications, these alloys are utilized in nuclear reactors and petrochemical equipment.

The Noble Metals

The noble or precious metals are a group of eight elements that have some physical characteristics in common. They are expensive and are superior or noble in properties—that is, characteristically soft, ductile, and oxidation resistant. The noble metals are silver, gold, platinum, palladium, rhodium, ruthenium, iridium, and osmium; the first three are most common and are used extensively in jewelry. Silver and gold may be strengthened by solid-solution alloying with copper; sterling silver is a silver-copper alloy containing approximately 7.5 wt% Cu. Alloys of both silver and gold are employed as dental restoration materials; also, some integrated circuit electrical contacts are of gold. Platinum is used for chemical laboratory equipment, as a catalyst (especially in the manufacture of gasoline), and in thermocouples to measure elevated temperatures.

Part 3 Basic Knowledge of Materials Chemistry

Vocabulary

extraction [ɪk'strækʃn]　　*n.* 提取；提炼；开采
refining [rɪ'faɪnɪŋ]　　*n.* 精炼；提纯；去除杂质
alloying [ə'lɔɪɪŋ]　　*n.* 合金化处理；炼制合金
fabrication [ˌfæbrɪ'keɪʃn]　　*n.* 制造，制作
ferrous ['ferəs]　　*adj.* 含铁的；铁的
versatile ['vɜːsətaɪl]　　*adj.* 多用途的；多功能的
unresponsive [ˌʌnrɪ'spɒnsɪv]　　*adj.* 无反应的
strengthening ['streŋθnɪŋ]　　*v.* 加强；巩固；强化
austenitizing ['ɔstinitaɪzɪŋ]　　*n.* 奥氏体化处理
quenching ['kwentʃɪŋ]　　*v.* 淬火
tempering ['tempərɪŋ]　　*v.* 回火
martensite ['mɑːtɪnzaɪt]　　*n.* 马氏体
martensitic [ˌmɑːtɪn'zɪtɪk]　　*adj.* 马氏体的
ferritic [fe'rɪtɪk]　　*adj.* 铁素体的
austenitic [ˌɔːstə'nɪtɪk]　　*adj.* 奥氏体的
cementite [sɪ'mentaɪt]　　*n.* 渗碳体
hexagonal [hek'sægənl]　　*adj.* 六边的，六角形的
elastic moduli [ɪ'læstɪk] ['mɒdjʊlaɪ]　　*n.* 弹性模量
centrifugal [ˌsentrɪ'fjuːgl]　　*adj.* 离心的
catalyst ['kætəlɪst]　　*n.* 催化剂

Lesson 13 Characteristics and Applications of Ceramics

Glass

The classification of ceramic materials on the basis of application is presented in Figure 13.1. The glasses are a familiar group of ceramics; containers, lenses, and fiberglass represent typical applications. As already mentioned, they are noncrystalline silicates containing other oxides, notably CaO, Na_2O, K_2O, and Al_2O_3, which influence the glass properties. A typical soda-lime glass consists of approximately 70 wt% SiO_2, the balance being mainly Na_2O and CaO. The two prime assets of these materials are their optical transparency and the relative ease with which they may be fabricated.

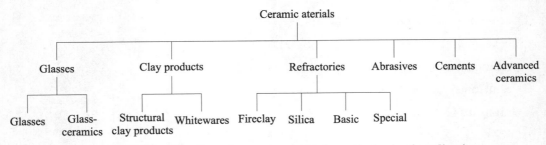

Figure 13.1 Classification of ceramic materials on the basis of application.

Clay products

One of the most widely used ceramic raw materials is clay. This inexpensive ingredient, found naturally in great abundance, often is used as mined without any upgrading of quality. Another reason for its popularity lies in the ease with which clay products may be formed; when mixed in the proper proportions, clay and water form a plastic mass that is very amenable to shaping. The formed piece is dried to remove some of the moisture, after which it is fired at an elevated temperature to improve its mechanical strength.

Most of the clay-based products fall within two broad classifications: the structural clay products and the whitewares. Structural clay products include building bricks, tiles, and sewer pipes-applications in which structural integrity is important. The whiteware ceramics become white after the high-temperature firing. Included in this group are porcelain, pottery, tableware, china, and sanitary ware. In addition to clay, many of these products also contain nonplastic ingredients, which influence the changes that take place during the drying and firing processes, and the characteristics of the finished products.

Refractories

Another important class of ceramics that are utilized in large tonnages is the refractory ceramics. The salient properties of these materials include the capacity to withstand high temperatures without melting or decomposing, and the capacity to remain unreactive and inert when exposed to severe environments. In addition, the ability to provide thermal insulation is often an important consideration. Refractory materials are marketed in a variety of forms, but bricks are the most common. Typical applications include furnace linings for metal refining, glass manufacturing, metallurgical heat treatment, and power generation.

Of course, the performance of a refractory ceramic, to a large degree, depends on its composition. On this basis, there are several classifications-namely, fireclay, silica, basic, and special refractories. For many commercial materials, the raw ingredients consist of both large (or grog) particles and fine particles, which may have different compositions. Upon firing, the fine particles normally are involved in the formation of a bonding phase, which is responsible for the increased strength of the brick; this phase may be predominantly either glassy or crystalline. The service temperature is normally below that at which the refractory piece was fired.

Cements

Several familiar ceramic materials are classified as inorganic cements: cement, plaster, and lime, which, as a group, are produced in extremely large quantities. The characteristic feature of these materials is that when mixed with water, they form a paste that subsequently sets and hardens. This trait is especially useful in that solid and rigid structures having just about any shape may be expeditiously formed. Also, some of these materials act as a bonding phase that chemically binds particulate aggregates into a single cohesive structure. Under these circumstances, the role of the cement is similar to that of the glassy bonding phase that forms when clay products and some refractory bricks are fired. One important difference, however, is that the cementitious bond develops at room temperature. Of this group of materials, portland cement is consumed in the largest tonnages.

Advanced ceramics

Although the traditional ceramics discussed previously account for the bulk of the production, the development of new and what are termed "advanced ceramics" has begun and will continue to establish a prominent niche in our advanced technologies. In particular, electrical, magnetic, and optical properties and property combinations unique to ceramics have been exploited in a host of new products. Furthermore, advanced ceramics are utilized in optical fiber communications systems, in microelectromechanical systems, as ball bearings, and in applications that exploit the piezoelectric behavior of a number of ceramic materials.

Vocabulary

porcelain ['pɔːsəlɪn]　*n.* 瓷器
pottery ['pɒtəri]　*n.* 陶器
tableware ['teɪblweə(r)]　*n.* 餐具
fireclay ['faɪəkleɪ]　*n.* 耐火土
cement [sɪ'ment]　*n.* 水泥
plaster ['plɑːstər]　*n.* 石膏
lime [laɪm]　*n.* 石灰
advanced ceramics [əd'vɑːnst] [sə'ræmɪks]　先进陶瓷

Part 3　Basic Knowledge of MATERIALS Chemistry

Lesson 14　Characteristics and Applications of Polymers

Plastics

Possibly the largest number of different polymeric materials come under the plastic classification. Plastics are materials that have some structural rigidity under load, and are used in general-purpose applications. Polyethylene, polypropylene, polyvinyl chloride, polystyrene, and the fluorocarbons, epoxies, phenolics, and polyesters may all be classified as plastics. They have a wide variety of combinations of properties. Some plastics are very rigid and brittle. Others are flexible, exhibiting both elastic and plastic deformations when stressed, and sometimes experiencing considerable deformation before fracture.

Polymers falling within this classification may have any degree of crystallinity, and all molecular structures and configurations (linear, branched, isotactic, etc.) are possible. Plastic materials may be either thermoplastic or thermosetting; in fact, this is the manner in which they are usually subclassified. However, to be considered plastics, linear or branched polymers must be used below their glass transition temperatures (if amorphous) or below their melting temperatures (if semicrystalline), or must be crosslinked enough to maintain their shape.

Elastomers

Natural rubber is still utilized to a large degree because it has an outstanding combination of desirable properties. However, the most important synthetic elastomer is styrene butadiene rubber(SBR), which is used predominantly in automobile tires, reinforced with carbon black. Nitrile butadiene rubber(NBR), which is highly resistant to degradation and swelling, is another common synthetic elastomer. The typical properties and applications of common elastomers depend on the degree of vulcanization and on whether any reinforcement is used.

For many applications (e.g., automobile tires), the mechanical properties of even vulcanized rubbers are not satisfactory in terms of tensile strength, abrasion and tear resistance, and stiffness. These characteristics may be further improved by additives such as carbon black. Finally, some mention should be made of the silicone rubbers. For these materials, the backbone chain is made of alternating silicon and oxygen atoms.

Fibers

The fiber polymers are capable of being drawn into long filaments having at least a 100 : 1 length-to-diameter ratio. Most commercial fiber polymers are utilized in the textile industry, being woven or knit into cloth or fabric. In addition, the aramid fibers are employed in composite materials. To be useful as a textile material, a fiber polymer must have a host of rather restrictive physical and chemical properties. While in use, fibers may be subjected to a variety of mechanical deformations (stretching, twisting, shearing, and abrasion). Consequently, they must have a high tensile strength (over a relatively wide temperature range) and a high modulus of elasticity, as well as abrasion resistance. These properties are governed by the chemistry of the polymer chains and also by the fiber drawing process.

The molecular weight of fiber materials should be relatively high or the molten material will be too weak and will break during the drawing process. Also, because the tensile strength increases with degree of crystallinity, the structure and configuration of the chains should allow the production of a highly crystalline polymer. That translates into a requirement for linear and unbranched chains that are symmetrical and have regular repeat units. Polar groups in the polymer also improve the fiber-forming properties by increasing both crystallinity and the intermolecular forces between the chains.

Coatings

Coatings are frequently applied to the surface of materials to serve one or more of the following functions: (1) to protect the item from the environment that may produce corrosive or deteriorative reactions; (2) to improve the item's appearance; and (3) to provide electrical insulation. Many of the ingredients in coating materials are polymers, the majority of which are organic in origin. These organic coatings fall into several different classifications, as follows: paint, varnish, enamel, lacquer, and shellac.

Many common coatings are latexes. A latex is a stable suspension of small insoluble polymer particles dispersed in water. These materials have become increasingly popular because they don't contain large quantities of organic solvents that are emitted into the environment—that is, they have low volatile organic compound (VOC) emissions. VOCs react in the atmosphere to produce smog. Large users of coatings such as automobile manufacturers continue to reduce their VOC emissions to comply with environmental regulations.

Part 3　Basic Knowledge of Materials Chemistry

Foams

Foams are plastic materials that contain a relatively high volume percentage of small pores and trapped gas bubbles. Both thermoplastic and thermosetting materials are used as foams; these include polyurethane, rubber, polystyrene, and poly(vinyl chloride). Foams are commonly used as cushions in automobiles and furniture as well as in packaging and thermal insulation. The foaming process is often carried out by incorporating into the batch of material a blowing agent that, upon heating, decomposes with the liberation of a gas. Gas bubbles are generated throughout the now-fluid mass, which remain in the solid upon cooling and give rise to a spongelike structure. The same effect is produced by dissolving an inert gas into a molten polymer under high pressure. When the pressure is rapidly reduced, the gas comes out of solution and forms bubbles and pores that remain in the solid as it cools.

Vocabulary

plastics ['plæstɪks]　*n.* 塑料

polyethylene [ˌpɒli'eθəliːn]　*n.* 聚乙烯

polypropylene [ˌpɒli'prəupəliːn]　*n.* 聚丙烯

polyvinyl chloride [ˌpɒlɪ'vaɪn(ə)l] ['klɔːraɪd]　聚氯乙烯

polystyrene [ˌpɒli'staɪriːn]　*n.* 聚苯乙烯

fluorocarbons [ˌfluroʊ'kɑːrbəns]　*n.* 碳氟化合物

polyester [ˌpɒli'estə]　*n.* 聚酯

thermoplastic [ˌθɜːməʊ'plæstɪk]　*adj.* 热塑性的

thermosetting ['θɜːməʊsetɪŋ]　*adj.* 热固性的

elastomer [ɪ'læstəmər]　*n.* 弹性体

butadiene [ˌbjutə'daɪin]　*n.* 丁二烯

rubber ['rʌbər]　*n.* 橡胶；合成橡胶

fiber ['faɪbər]　*n.* 纤维

aramid ['ærəmɪd]　*n.* 芳族聚酰胺

elasticity [ˌiːlæ'stɪsəti]　*n.* 弹性；弹力

crystallinity [ˌkrɪstə'lɪnəti]　*n.* 结晶度

paint [peɪnt]　*n.* 油漆

varnish ['vɑːnɪʃ]　*n.* 清漆，亮光漆

enamel [ɪ'næml]　　*n.* 搪瓷

lacquer ['lækər]　　*n.* 漆器

shellac [ʃə'læk]　　*n.* 虫胶

latex ['leɪteks]　　*n.* 乳胶；乳液

foam [fəʊm]　　*n.* 泡沫

polyurethane [ˌpɒli'jʊərəθeɪn]　　*n.* 聚氨酯

Part 3 Basic Knowledge of MATERIALS Chemistry

Lesson 15 Characteristics and Applications of Composites

Polymer-Matrix Composites

Polymer-matrix composites consist of a polymer resin as the matrix, with fibers as the reinforcement medium. These materials are used in the greatest diversity of composite applications, as well as in the largest quantities, in light of their room-temperature properties, ease of fabrication, and cost. In this section the various classifications of polymer-matrix composites are discussed according to reinforcement type (i.e., glass, carbon, and aramid), along with their applications and the various polymer resins that are employed.

Glass Fiber-Reinforced Polymer Composites

Fiberglass is simply a composite consisting of glass fibers, either continuous or discontinuous, contained within a polymer matrix; this type of composite is produced in the largest quantities. The fiber diameters normally range between 3 m and 20 m. Many fiberglass applications are familiar: automotive and marine bodies, plastic pipes, storage containers, and industrial floorings. The transportation industries are utilizing increasing amounts of glass fiber-reinforced plastics in an effort to decrease vehicle weight and boost fuel efficiencies. A host of new applications are being used or currently investigated by the automotive industry.

Carbon Fiber-Reinforced Polymer Composites

Carbon is a high-performance fiber material that is the most commonly used reinforcement in advanced (i.e., nonfiberglass) polymer-matrix composites. Carbon fibers have the highest specific modulus and specific strength of all reinforcing fiber materials. These fibers exhibit a diversity of physical and mechanical characteristics, allowing composites incorporating these fibers to have specific engineered properties. Fiber and composite manufacturing processes have been developed that are relatively inexpensive and cost effective. Carbon-reinforced polymer composites are currently being utilized extensively in sports and recreational equipment (fishing rods, golf clubs), filament-wound rocket motor cases, pressure vessels, and aircraft structural components—both military and commercial, fixed wing and helicopters (e.g., as wing, body, stabilizer, and rudder components).

Aramid Fiber-Reinforced Polymer Composites

Aramid fibers are high-strength, high-modulus materials that were introduced in the early 1970s. They are especially desirable for their outstanding strength-to weight ratios, which are superior to metals. Chemically, this group of materials is known as poly(paraphenylene terephthalamide). There are a number of aramid materials; trade names for two of the most common are Kevlar and Nome. For the former, there are several grades (viz. Kevlar 29, 49, and 149) that have different mechanical behaviors. Mechanically, these fibers have longitudinal tensile strengths and tensile moduli that are higher than other polymeric fiber materials; however, they are relatively weak in compression. In addition, this material is known for its toughness, impact resistance, and resistance to creep and fatigue failure.

The aramid fibers are most often used in composites having polymer matrices; common matrix materials are the epoxies and polyesters. Since the fibers are relatively flexible and somewhat ductile, they may be processed by most common textile operations.Typical applications of these aramid composites are in ballistic products (bulletproof vests and armor), sporting goods, tires, ropes, missile cases, pressure vessels, and as a replacement for asbestos in automotive brake and clutch linings, and gaskets.

Other Fiber Reinforcement Materials

Other fiber materials that are used to much lesser degrees are boron, silicon carbide, and aluminum oxide. Boron fiber-reinforced polymer composites have been used in military aircraft components, helicopter rotor blades, and some sporting goods. Silicon carbide and aluminum oxide fibers are utilized in tennis rackets, circuit boards, military armor, and rocket nose cones.

Metal-matrix composites

As the name implies, for metal-matrix composites the matrix is a ductile metal. These materials may be utilized at higher service temperatures than their base metal counterparts; furthermore, the reinforcement may improve specific stiffness, specific strength, abrasion resistance, creep resistance, thermal conductivity, and dimensional stability. Some of the advantages of these materials over the polymer-matrix composites include higher operating temperatures, nonflammability, and greater resistance to degradation by organic fluids. Metal-matrix composites are much more expensive than polymer-matrix composites, and therefore, their use is somewhat restricted.

Automobile manufacturers have recently begun to use metal-matrix composites in their products. For example, some engine components have been introduced consisting of an aluminum-alloy matrix that is reinforced with aluminum oxide and carbon fibers; this metal-matrix composite is light in weight and resists wear and thermal distortion. Metal-matrix composites are also employed in driveshafts (that have higher rotational speeds and reduced vibrational noise levels), extruded stabilizer bars, and forged suspension and transmission components. The aerospace industry also uses metal-matrix composites. Structural applications include advanced aluminum alloy metal-matrix composites; boron fibers are used as the reinforcement for the Space Shuttle Orbiter, and continuous graphite fibers for the Hubble Telescope.

Ceramic-matrix composites

Ceramic materials are inherently resilient to oxidation and deterioration at elevated temperatures; were it not for their disposition to brittle fracture, some of these materials would be ideal candidates for use in high-temperature and severe-stress applications, specifically for components in automobile and aircraft gas turbine engines. The fracture toughnesses of ceramics have been improved significantly by the development of a new generation of ceramic-matrix composites-particulates, fibers, or whiskers of one ceramic material that have been embedded into a matrix of another ceramic.

Ceramic-matrix composites may be fabricated using hot pressing, hot isostatic pressing, and liquid phase sintering techniques. Relative to applications, SiC whisker-reinforced aluminas are being utilized as cutting tool inserts for machining hard metal alloys; tool lives for these materials are greater than for cemented carbides.

Carbon-carbon composites

One of the most advanced and promising engineering material is the carbon fiber-reinforced carbon-matrix composite, often termed a carbon-carbon composite; as the name implies, both reinforcement and matrix are carbon. These materials are relatively new and expensive and, therefore, are not currently being utilized extensively. Their desirable properties include high-tensile moduli and tensile strengths that are retained to temperatures in excess of resistance to creep, and relatively large fracture toughness values. Furthermore, carbon-carbon composites have low coefficients of thermal expansion and relatively high thermal conductivities; these characteristics, coupled with high strengths, give rise to a relatively low susceptibility to thermal shock. Their major drawback is a propensity to high temperature oxidation.

The carbon-carbon composites are employed in rocket motors, as friction materials in aircraft and high performance automobiles, for hot-pressing molds, in components for advanced turbine engines, and as ablative shields for re-entry vehicles.

Hybrid composites

A relatively new fiber-reinforced composite is the hybrid, which is obtained by using two or more different kinds of fibers in a single matrix; hybrids have a better all around combination of properties than composites containing only a single fiber type. A variety of fiber combinations and matrix materials are used, but in the most common system, both carbon and glass fibers are incorporated into a polymeric resin. The carbon fibers are strong and relatively stiff and provide a low-density reinforcement; however, they are expensive. Glass fibers are inexpensive and lack the stiffness of carbon. The glass-carbon hybrid is stronger and tougher, has a higher impact resistance, and may be produced at a lower cost than either of the comparable all-carbon or all-glass reinforced plastics.

Principal applications for hybrid composites are lightweight land, water, and air transport structural components, sporting goods, and lightweight orthopedic components.

Vocabulary

composite ['kɒmpəzɪt]　　*n.* 复合材料

reinforcement [ˌriːɪnˈfɔːsmənt]　　*n.* 增强体

poly(paraphenylene terephthalamide) [pɔliˈpapəfiːniliːn] [ˌterefˈθæleɪt]　　*n.* 聚对苯二甲酰对苯二胺

nonflammability [ˈnɒnflæməˈbɪləti]　　*n.* 不燃性

toughness [ˈtʌfnəs]　　*n.* 韧性

pressing [ˈpresɪŋ]　　*n.* 挤压；冲压

sintering [ˈsɪntərɪŋ]　　*n.* 烧结

whisker [ˈwɪskər]　　*n.* 晶须

coefficient [ˌkəʊɪˈfɪʃnt]　　*n.* 系数

hybrid [ˈhaɪbrɪd]　　*n.* 杂化物；杂化材料

Part 4

Typical Chemical Process

Lesson 16 Operations in Chemical Engineering

Before reading this unit, try to answer the following questions:
1. Is chemical reaction involved in unit operations?
2. Who first presented the concept of operations clearly?
3. How many kinds of unit operations can you list?
4. Can you identify two major physical models: ideal contact and rate of transfer?

Chemical processes may consist of widely varying sequences of steps, the principles of which are independent of the material being operated upon and of other characteristics of the particular system. In the design of a process, each step to used can be studied individually if the step sister are recognized. Some of the steps are chemical reactions, whereas others are physical changes. The versatility of chemical engineering originates in training to the practice of breaking up a complex process into individual physical steps, called unit operations, and into the chemical reactions. The unit-operations concept in chemical engineering is based on the philosophy that the widely varying sequences of steps can be reduced to simple operations or reactions, which are identical in fundamentals regardless of the material being processed. This principle, which became obvious to the pioneers during the development of the American chemical industry, was first clearly presented by A. D. Little in 1915.

Any chemical process, on whatever scale conducted, may be resolved into a coordinated series of what may be termed "unit actions", as pulverizing, mixing heating, roasting, absorbing, condensing, lixiviating, precipitating, crystallizing, filtering, dissolving, electrolyzing and so on. The number of these basic unit operations is not very large and relatively few of them are involved in any particular process. The complexity of chemical engineering results from the variety of conditions as to temperature, pressure, etc., under which the unit actions must be carried out in different processes and from the limitations as to materials of construction and design of apparatus imposed by the physical and chemical character of the reacting substances.

The original listing of the unit operations quoted above names twelve actions, not all of which are considered unit operations. Additional ones have been designated since then, at a modest rate over the years but recently at an accelerating rate. Fluid flow, heat transfer, distillation, humidification, gas absorption, sedimentation, classification, agitation, and centrifugation have long been recognized. In recent years increasing understanding of new techniques and adaptation of old but seldom used separative techniques has led to a continually increasing number of separations, processing operations, or steps in a manufacture that could be used without significant alteration in variety of processes. This is the basis of a terminology of "unit operations", which now offers us a list of techniques.

Classification of Unit Operations

(1) Fluid flow. This concerns the principles that determine the flow or transportation of any fluid from one point to another.

(2) Heat transfer. This unit operation deals with the principles that government accumulation and transfer of heat and energy from one place to another.

(3) Evaporation. This is a special case of heat transfer, which deals with the evaporation of a volatile solvent such as water from a nonvolatile solute such as salt or any other material in solution.

(4) Drying. In this operation volatile liquids, usually water, are removed from solid materials.

(5) Distillation. This is an operation whereby components of a liquid mixture are separated by boiling because of their differences in vapor pressure.

(6) Absorption. In this process a component is removed from a gas stream by treatment with a liquid.

(7) Membrane separation. This process involves the diffusion of a solute from a liquid or gas through a semipermeable membrane barrier to another fluid.

(8) Liquid-liquid extraction. In this case a solute in a liquid solution is removed by contacting with another liquid solvent which is relatively immiscible with the solution.

(9) Liquid-solid leaching. This involves treating a finely divided solid with a liquid that dissolves out and removes a solute contained in the solid.

(10) Crystallization. This concerns the removal of a solute such as a salt from a solution by precipitating the solute from the solution.

(11) Mechanical physical separations. These involve separation of solids, liquids, or gases by mechanical means, such as filtration, setting, and size reduction, which are often classified as separate unit operations.

Many of these unit operations have certain fundamental and basic principles or mechanisms in common. For example, the mechanism of diffusion or mass transfer occurs in drying, absorption, distillation, and crystallization. Heat transfer occurs in drying, distillation, evaporation, and so on.

Fundamental Concepts

Because the unit operations are a branch of engineering, they are based on both science and experience. Theory and practice must combine to yield designs for equipment that can be fabricated, assembled, operated, and maintained. The following four concepts are basic and form the foundation for the calculation of all operations.

The Material Balance

If matter may be neither created nor destroyed, the total mass for all materials entering an operation equals the total mass for all materials leaving that operation, except for any material that may be retained or accumulated in the operation. By the application of this principle, the yields of a chemical reaction or engineering operation are computed.

In continuous operations, material is usually not accumulated in the operation, and a material balance consists simply in charging (or debiting) the operation with all material entering and crediting the operation with all material leaving the same manner as used by any accountant. The result must be a balance.

As long as the reaction is chemical and does not destroy or create atoms, it is proper and frequently very convenient to employ atoms as the basis for the material balance. The material balance may be made for the entire plant or for any part of it as a unit, depending upon the problem at hang.

The Energy Balance

Similarly, an energy balance may be made around any plant or unit operation to determine the energy required to carry on the operation or to maintain the desired operating conditions. The principle is just as important as that of the material balance, and it is used in the same way. The important point to keep in mind is that all energy of all kinds must be included, although it may be converted to a single equivalent form.

The Ideal Contact (The Equilibrium Stage Model)

Whenever the materials being processed are in contact for any length of time under specified conditions, such as conditions of temperature, pressure, chemical composition, or electrical potential they tend to approach a definite condition of equilibrium which is determined by the specified conditions. In many cases the rate of approach to these equilibrium conditions is so rapid or the length of time is sufficient that the equilibrium conditions are practically attained at each contact. Such a contact is known as an equilibrium or ideal contacts.

The calculation of the number of ideal contacts is an important step required in understanding those unit operations involving transfer of material from one phase to another, such as leaching, extraction, absorption, and dissolution.

Rates of an Operation (The Rate of Transfer Model)

In most operations equilibrium is not attained either because of insufficient time or because it is not desired. As soon as equilibrium is attained no further change can take place and the process stops, but the engineer must keep the process going. For this reason rate operations, such as rate of energy transfer, rate of mass transfer, and rate of chemical reaction, are of the greatest importance and interest. In all such cases the rate and direction depend upon a difference in potential or driving force. The rate usually may be expressed as proportional to a potential drop divided by a resistance. An application of this principle to electrical energy is the familiar Ohm's law for steady or direct current.

In solving rate problems as in heat transfer or mass transfer with this simple concept, the major difficulty is the evaluation of the resistance terms are generally computed from an empirical correlation of many determinations of transfer rates under different conditions.

The basic concept that rate depends directly upon a potential drop and inversely upon a resistance may be applied to any rate operation, although the rate may be expressed in different ways with particular coefficients for particular cases.

Vocabulary

lixiviate [lik'sivieit]   *vt.* 浸提（析，出）；溶滤

dissolve [di'zɔlv]   *v.*；*n.* 使溶解；溶化

humidification [hju:ˌmidifi'keiʃən]   *n.* 增湿作用；湿润

sedimentation [sedimen'teiʃən]   *n.* 沉积；沉淀；沉降；淀积

semipermeable ['semi'pə:miəbl]   *adj.* 半渗透性的

immiscible [i'misəbl]　*adj.* 不混溶的；不互溶的

leaching ['li:tʃiŋ]　*n.* 浸取；浸提

solute ['səlju:t]　*n.* 溶质；溶解物

settling ['setliŋ]　*n.* 沉降；沉淀

debit ['debit]　*vt; n.*（记入）借方

equilibrium stage　平衡级

dissolution [disə'lju:ʃən]　*n.* 溶解；溶化

correlation [kɔri'leiʃən]　*n.* 相关（性）；（相互，对比）关系

Lesson 17 Distillation

Before reading this unit, try to answer the following questions:
1. What does the distillation process primarily depend on?
2. How is the equilibrium stage concept used in the distillation design?
3. Can you list several kinds of plates and packings? What are their features?
4. When should we utilize batch distillation in stead of continuous one?

Separation operations achieve their objective by the creation of two or more coexisting zones which differ in temperature, pressure, composition, and/or phase state. Each molecular species in the mixture to be separated reacts in a unique way to differing environments offered by these zones. Consequently, as the system moves toward equilibrium, each establishes a different concentration in each zone, and this results in a separation between the species.

The separation operation called distillation utilizes vapor and liquid phases at essentially the same temperature and pressure for the coexisting zones. Various kinds of device such as dumped or ordered packings and plates or trays are used to bring the two phases into intimate contact. Trays are stacked one above the other and enclosed in a cylindrical shell to form a column. Packings are also generally contained in a cylindrical shell between hold-down and support plates.

Continuous Distillation

The feed material, which is to be separated into fractions, is introduced at one or more points along the column shell. Because of the difference in gravity between vapor and liquid phases, liquid runs down the column, cascading from tray to tray, while vapor flows up the column, contacting liquid at each tray.

Liquid reaching the bottom of the column is partially vaporized in a heated reboiler to provide boil-up, which is sent back up the column. The remainder of the bottom liquid is withdrawn as bottoms, or bottom product. Vapor reaching the top of the column is cooled and condensed to liquid in the overhead condenser. Part of this liquid is returned to the column as reflux to provide liquid overflow. The remainder of the overhead stream is withdrawn as distillate, or overhead product.

This overall flow pattern in a distillation column provides countercurrent contacting of vapor and liquid streams on all the trays through the column. Vapor and liquid phases on a given tray approach thermal, pressure, and composition equilibriums to an extent dependent upon the efficiency of the contacting tray.

The lighter (lower-boiling) components tend to concentrate in the vapor phase, while the heavier (higher-boiling) components tend toward the liquid phase. The result is a vapor phase that becomes richer in light components as it passes up the column and a liquid phase that becomes richer in heavy components as it cascades downward. The overall separation achieved between the distillate and the bottoms depends primarily on the relative volatilities of the components, the number of contacting trays, and the ratio of the liquid-phase flow rate to the vapor-phase flow rat.

If the feed is introduced at one point along the column shell, the column is divided into an upper section, which is often called the rectifying section, and a lower section, which is often referred to as the stripping section. These terms become rather indefinite in multiple-feed columns and columns from which a product sidestream is withdrawn somewhere along the column length in addition to the two end-product streams.

Equilibrium-Stage Concept

Energy and mass-transfer processes in an actual distillation column are much too complicated to be readily modeled in any direct way. This difficulty is circumvented by the equilibrium-stage model, in which vapor and liquid streams leaving an equilibrium stage are in complete equilibrium with each other and thermodynamic relations can be used to determine the temperature of and relate the concentrations in the equilibrium streams at a given pressure. A hypothetical column composed of equilibrium stages (instead of actual contact trays) is designed to accomplish the separation specified for the actual column. The number of hypothetical equilibrium stages required is then converted to a number of actual trays by means of tray efficiencies, which describe the extent to which the performance of an actual contact tray duplicates the performance of an equilibrium stage.

Use of the equilibrium-stage concept separates the design of a distillation column into three major steps: (1) Thermodynamic data and methods needed to predict equilibrium-phase compositions are assembled. (2) The number of equilibrium stages required to accomplish a specified separation, or the separation that will be accomplished in a given number of equilibrium stages, is calculated. (3) The number of equilibrium stages is converted to an equivalent number of actual contact trays or height of packing, and the column diameter is determined.

All separation operations require energy input in the form of heat or work. In the conventional distillation operation, energy required to separate the species is added in the form of heat to the reboiler at the bottom of the column, where the temperature is highest. Also, heat is removed from a condenser at the top of the column, where the temperature is lowest. This frequently results in a large energy-input requirement and low overall thermodynamic efficiency. With recent dramatic increases in energy costs, complex distillation operations that offer higher thermodynamic efficiency and lower energy-input requirements are being explored.

Related Separation Operations

The simple and complex distillation operations just described all have two things in common: (1) both rectifying and stripping sections are provided so that a separation can be achieved between two components that are adjacent in volatility; and (2) the separation is effected only by the addition and removal of energy and not by the addition of any mass separating agent (MSA) such as in liquid-liquid extraction. Sometimes, alternative single- or multiple-stage vapor-liquid separation operations may be more suitable than distillation for the specified task.

Batch Distillation

Batch distillation, which is the process of separating a specific quantity (the charge) of a liquid mixture into products, is used extensively in the laboratory and in small production units that may have to serve for many mixtures. When there are N components in the feed, one batch column will suffice where $N-1$ simple continuous distillation columns would be required.

Many larger installations also feature a batch still. Material to be separated may be high in solids content, or it might contain tars or resins that would plug or foul a continuous unit. Use of a batch unit can keep solids separated and permit convenient removal at the termination of the process.

Simple Batch Distillation

The simplest form of batch still consists of a heated vessel (pot or boiler), a condenser, and one or more receiving tanks. No trays or packing are provided. Feed is charged into the vessel and brought to boiling. Vapors are condensed and collected in a receiver. No reflux is returned. The rate of vaporization is sometimes controlled to prevent "bumping" the charge and to avoid overloading the condenser, but other controls are minimal. This process is often referred to as Rayleigh distillation.

The simple batch still provides only one theoretical plate of separation. Its use is usually restricted to preliminary work in which products will be held for additional separation at a later time, when most of the volatile component must be removed from the batch before it is processed further, or for similar noncritical separations.

Batch Distillation with Rectification

To obtain products with a narrow composition range, a rectifying batch still is used that consists of a pot (or reboiler), a rectifying column, a condenser, some means of splitting off a portion of the condensed vapor (distillate) as reflux, and one or more receivers. Temperature of the distillate is controlled in order to return the reflux at or near the column temperature to permit a true indication of reflux quantity and to improve column operation. The column may also operate at elevated pressure or vacuum, in which case appropriate device must be included to obtain the desired pressure. Equipment design methods for batch-still components, except for the pot, follow the same principles as those presented for continuous units, but the design should be checked for each mixture if several mixtures are to be processed. It should also be checked at more than one point of a mixture, since composition in the column changes as distillation proceeds. Pot design is based on batch size and required vaporization rate.

In operation, a batch of liquid is charged to the pot and the system is first brought to steady state under total reflux. A portion of the overhead condensate is then continuously withdrawn in accordance with the established reflux policy. Cuts are made by switching to alternate receivers, at which time operating conditions may be altered. The entire column operates as an enriching section. As time proceeds, composition of the material being distilled becomes less rich in the more volatile components, and distillation of a cut is stopped when accumulated distillate attains the desired average composition.

The progress of batch distillation can be controlled in several ways:

(1) Constant reflux, varying overhead composition. Reflux is set at a predetermined value at which it is maintained for the run. Since pot liquid composition is changing, instantaneous composition of the distillate also changes. Distillation is continued until the average distillate composition is at the desired value. In the case of a binary, the overhead is then diverted to another receiver, and an intermediate cut is withdrawn until the remaining pot liquor meets the required specification. The intermediate cut is usually added to the next batch. For a multicomponent mixture, two or more intermediate cuts may be taken between product cuts.

(2) Constant overhead composition, varying reflux. If it is desired to maintain a constant overhead composition in the case of a binary, the amount of reflux returned to the columns must be constantly increased throughout the run. As time proceeds, the pot is gradually depleted of the lighter component. Finally, a point is reached at which the reflux ratio has attained a very high value. The receivers are then changed, the reflux is reduced, and an intermediate cut is taken as before. This technique can also be extended to a multicomponent mixture.

(3) Other control methods. A cycling procedure can be used to set the pattern for column operation. The unit operates at total reflux until equilibrium is established. Distillate is then taken as total draw-off for a short period of time, after which the column is again returned to total-reflux operation. This cycle is repeated through the course of distillation. Another possibility is to optimize the reflux ratio in order to achieve the desired separation in a minimum of time. Complex operations may involve withdrawal of sidestreams, provision for intercondensers, addition of feeds to trays, and periodic charge addition to the pot.

Vocabulary

dumped packing　乱堆填料

ordered packing　整砌填料；规整填料

hold-down　*n.* 压具（板，块）；压紧（装置）；固定

feed [fi:d]　*n.* 进料，加料；加工原料

cascade [kaes'keid]　*v.*; *n.* 梯流，阶流式布置；级联，串级

boil-up　蒸出（蒸汽）

bottom ['bɔtəm]　*n.* (*pl.*) 底部沉淀物，残留物，残渣

relative volatility　相对挥发度（性）

rectify ['rektifai]　*vt.* 精馏，精炼，蒸馏

rectifying section　精馏段

stripping　*n.* 洗提；汽提；解吸

stripping section　提馏段

multiple-feed　多口进料

sidestream　*n.* 侧线馏分，塔侧抽出物

circumvent [,sɜ:kəm'vent]　*vt.* 绕过，回避，胜过

hypothetical [ˌhaipə'θetikəl]　*adj.* 假定（设，说）的；有前提的

duplicate ['djuːplikeit]　*vt.* 重复，加倍，复制

mass separating agent　质量分离剂

flash drum　闪蒸槽

rectifier ['rektifaiə]　*n.* 精馏器（塔）；整流器

bump [bʌmp]　*v.* 扰动，暴沸；冲击，造成凹凸

condensate [kən'denseit]　*n.* 冷凝物，冷凝液

　　　　　　　　　　　v. 冷凝，凝结

binary ['bainəri]　*adj.*; *n.* 二，二元的

deplete [di'pliːt]　*vt.* 放空，耗尽，使枯竭；贫化，减少

Lesson 18 Wet-Chemical Synthesis of Inorganic Nanocrystals

Numerous studies have focused on the development of synthesis methodologies to produce inorganic nanocrystals with tunable size and shape since Faraday successfully did a pioneering synthesis of colloidal gold in the 1850s. Divided into top-down and bottom-up, tremendous methods such as chemical vapor deposition, sputtering, and thermal decomposition have been well studied and established to prepare nanocrystals. Wet-chemical synthesis is one of the most efficient and convenient methods to prepare shape- and size controllable nanocrystals. With bench-top chemicals and pieces of glassware, wet-chemical synthesis can be easily operated regardless of any expensive, complicated, and dedicated instruments and equipment which need regular maintenance. Furthermore, it is one of the few methods that can be scaled up and applied in industrial manufacturing because of its convenience and better cost-effectiveness. As a result, wet-chemical synthesis is the most effective and popular way to produce shape-controlled nanocrystals.

Hydrothermal Synthesis/Solvothermal Synthesis

Hydrothermal synthesis is a process that utilizes a single heterogeneous phase reactions in aqueous media at elevated temperature and pressure to crystallize and make inorganic nanocrystals directly from solutions. This synthesis offers a low-temperature, direct route to inorganic powder with a narrow size distribution avoiding the calcination step. The mechanism of hydrothermal reaction follows a liquid nucleation model. Detailed principles are comprised of theories of chemical equilibrium, chemical kinetics, and thermodynamic properties of aqueous systems under hydrothermal conditions.

Applied to inorganic nanocrystals, the process involves heating metal salts, oxides, or hydroxides as a solution of suspension in liquid at controlled temperature and pressure for about 20 hours. Typically the temperature in a hydrothermal process falls between the boiling point of water and the critical temperature (T_c = 374 °C), while the pressure is over 100 kPa. The product is washed by deionized water to get rid of ions in the solvent and other impurities. After drying in air, fairly well dispersible inorganic nanocrystals are obtained.

The stability and particle size of the final product will depend on pH value, reagent concentrations, reaction temperature, and time. Since the mechanisms for hydrothermal reactions are different, the corresponding process conditions could vary greatly. Moreover, the hydrothermal synthesis can be enhanced, by hybridizing it with microwaves, electrochemistry, ultrasound, mechanochemistry, and optical radiation.

Sol-Gel Method

This method offers specific advantages in preparations of inorganic nanocrystals. The early formation of a gel provides a high degree of homogeneity and reduces the need of atomic diffusion during the solid state calcinations. Moreover, the processing often starts with metal alkoxides, many of which are liquids or volatile solids that can easily be purified, providing extremely pure oxide precursors. However, the relative high costs of the metal alkoxides may be prohibitive for certain applications and the release of large amounts of alcohol during the calcination step requires special safety considerations.

A solution of the appropriate precursors (metal salts of metal organic compounds) is formed first, followed by conversion into homogeneous oxide (gel) after hydrolysis and condensation. Drying and subsequent calcination of the gel yields an oxide product. Usually, for preparation of multi-component oxides, alkoxides are mixed together in alcohol. Components for which no alkoxides are available are introduced as salts, such as acetates. Hydrolysis is carried out under controlled temperature, pH, and concentration of alkoxides, added water, and alcohol.

Microemulsion Technique

The technique of chemical reactions in microemulsions to produce nanoparticles has already 20 years of history behind, but the mechanisms to control the final size and the size distribution are still not well known. The knowledge of the mechanism is a crucial step in order to extend the potential applications of this technique. Nowadays there is a great interest in nanotechnologies and the developing of simple and reproducible methods to synthesize nanomaterials has attracted the interest of many researchers. The microemulsion method is a good candidate for this purpose.

Microemulsions are thermodynamically stable systems composed of two inmiscible liquids (usually, water and oil) and a surfactant. Droplets of water-in-oil (W/O) or oil-in-water (O/W) are stabilized by surfactants when small amounts of water or oil are used, respectively. The size of the droplets can be controlled very precisely just by changing the ratio R = [water or oil] / [surfactant] in the nanometer range. These nanodroplets can be used as nanoreactors to carry out chemical reactions. It was initially assumed that these nanodroplets could be used as templates to control the final size of the particles, however, the research carried out in the last years has shown that besides the droplet size, several other parameters play an important role in the final size distribution.

The main idea behind the microemulsion technique is very simple: two reactants (A and B) are introduced in two identical microemulsions. After mixing both microemulsions, droplets collide and interchange the reactants. Then the reaction can take place inside the nanoreactors. Different variations of this procedure have been employed. For example, one of the reactants can be introduced in solution into a microemulsion carrying the other reactant; or can be added directly to the microemulsions as a solid, liquid or gas. The most commonly used method of synthesis is the use of two similar microemulsions containing the reactants.

Precipitation and Coprecipitation Approach

One of the oldest techniques for the preparation of inorganic nanocrystals is the precipitation of products from solutions. In precipitation reactions, the metal precursors are dissolved in an ordinary solvent, such as water, and a precipitating agent is added to generate an insoluble solid. The main advantage of precipitation reactions is that large quantities of particles can be obtained. Uniform particles are usually synthesized by a homogeneous precipitation reaction, a process that includes the separation of the nucleation and growth of the nuclei.

Coprecipitation is a facile and convenient approach to prepare inorganic nanocrystals. Since 1981 where Massart reported the synthesis of magnetic nanocrystals in acid and alkaline media to date this approach still used to obtain magnetic nanocrystals, especially, iron oxide. This technique consists on reducing a mixture of metallic ions (e.g., Fe^{2+} and Fe^{3+}) using a basic solution [usually, NaOH, NH_4OH, or $N(CH_3)_4OH$] at temperature below 100 °C. The advantages of the coprecipitation method are the high yield, high product purity, the lack of necessity to use organic solvents, easily reproducible, and low cost. However, the properties of the obtained particles, such as size, shape, and composition are highly dependent on the reaction parameters (temperature, pH, ionic strength, kind of basic solution, and so on). Homogeneous precipitation can be obtained via a process that involves separation of the nucleation and growth of the nuclei. Two stages are involved in this process: (1) a short burst of nucleation when the concentration of the species reaches critical supersaturation; and (2) a slow growth of the nuclei by diffusion of the solutes to the surface on the crystal. To obtain mono-dispersed nanoparticles, these two stages should ideally be separated, that is, nucleation should be avoided during the period of growth.

Vocabulary

wet-chemical 湿化学的

synthesis ['sɪnθəsɪs] *n.* 合成

nanocrystal *n.* 纳米晶体

methodology [ˌmeθə'dɒlədʒi] *n.* 方法论；（从事某一活动的）方法，原则

chemical vapor deposition 化学气相沉积（缩写词为 CVD）

sputtering *n.* 反应溅射法

thermal decomposition 热分解法

shape-controlled 形貌控制

Hydrothermal [ˌhaɪdrəʊ'θɜːməl] *adj.* 水热法的；水热的

heterogeneous [ˌhetərə'dʒiːniəs] *adj.* 非均匀的

crystallize ['krɪstəlaɪz] *v.* 结晶；（使）形成晶体

calcination *n.* 煅烧，焙烧

nucleation [ˌnukli'eɪʃən] *n.* 成核现象，晶核形成

chemical equilibrium 化学平衡

chemical kinetics 化学动力学

thermodynamic *adj.* 热力学的，使用热动力的

dispersible *adj.* 可分散的；分散片；可分散；分散

reagent [ri'eɪdʒənt] *n.* 反应物，试剂

concentration [ˌkɒnsn'treɪʃn] *n.* 浓度

mechanochemistry *n.* 力化学，机械化学；机械化学现象

preparation [ˌprepə'reɪʃn] *n.* 制备

homogeneity [ˌhɒmədʒə'niːəti] *n.* 同质性；同质；同种

diffusion [dɪ'fjuːʒən] *n.* 扩散；传播；漫射

alkoxide *n.* 醇盐；（pl.）烃氧化物类

precursor [pri'kɜːsə(r)] *n.* 前驱体

homogeneous [ˌhɒmə'dʒiːniəs] *adj.* 同种类的；均匀的

hydrolysis [haɪ'drɒlɪsɪs] *n.* 水解

condensation [ˌkɒnden'seɪʃn] *n.* 凝结

microemulsion *n.* 微乳液

surfactant [sɜː'fæktənt] *n.* 表面活性剂

template ['templeɪt]   n. 模板

parameter [pə'ræmɪtə(r)]   n. 参数

precipitation [prɪˌsɪpɪ'teɪʃn]   n. 沉淀

coprecipitation   n. 共沉淀

insoluble [ɪn'sɒljəbl]   adj. 不溶的

nuclei ['njuːklɪaɪ]   n. 晶核

References

[1] TIMBERLAKE K C. Chemistry: an introduction to general, organic, and biological chemistry[M]. 13th edition. London: Pearson, 2015.

[2] GEANKOPLES C. Transport processes and unit operations[M]. 2nd Edition. Boston: Allyn and Bacon, 1983.

[3] WALAS S M. Chemical process equipment[M]. Oxford: Butterworth Publishers, 1988.

[4] 马永祥，孙晓君. 化学专业英语[M]. 兰州：兰州大学出版社，2019.